Robert Henry Scott

Weather Charts and Storm Warnings

Robert Henry Scott

Weather Charts and Storm Warnings

ISBN/EAN: 9783337336967

Printed in Europe, USA, Canada, Australia, Japan

Cover: Foto ©berggeist007 / pixelio.de

More available books at **www.hansebooks.com**

WEATHER CHARTS

AND STORM WARNINGS

BY

ROBERT H. SCOTT, M.A., F.R.S.

DIRECTOR OF THE METEOROLOGICAL OFFICE

WITH NUMEROUS ILLUSTRATIONS

HENRY S. KING & CO., LONDON
1876

PREFACE.

THIS little work has been put together in the endeavour to supply a want which has been expressed in many quarters: that of an explanation of the weather charts which appear in the newspapers, and of the remarks which are appended to them.

The conceptions and principles on which the science of weather study is based are apparently quite new to the majority of ordinary readers, who still hold to the belief that the barometer rises or falls in direct relation to the weather, without any attempt to consider *how* or *why* it does so.

It is hoped that the following pages may convey some idea, however imperfect, of the present state of Weather Knowledge, as distinguished from the science of Meteorology itself, of which this book makes no claim to be called a manual.

With a very few exceptions, theories of the causes of storms have been left unnoticed, as the object has been to explain to the reader what he can learn from a careful study of the information published in the newspapers or in the daily weather reports, and therefore accessible to all.

In treating of a science now in process of rapid development, it can only be expected that every year will add to our knowledge, and that many of the principles stated in these pages will be extended or modified by the results of subsequent experience. All that it is here attempted to give to the public is an account of the actual state of our knowledge at present.

I must express my sincere thanks to several friends who have aided me with their advice during the preparation of the book, and most notably to Mr. Frederic Gaster, of this Office.

The illustrations which are contained in the text have all been executed by the Patent Type Founding Company, by the same process as they employ for the production of charts for newspapers.

ROBERT H. SCOTT.

METEOROLOGICAL OFFICE:
June 10, 1876.

CONTENTS.

CHAPTER		PAGE
I.	THE MATERIALS AVAILABLE FOR WEATHER STUDY	1
II.	THE WIND	19
III.	THE BAROMETER	28
IV.	GRADIENTS	40
V.	CYCLONES AND ANTICYCLONES	53
VI.	THE MOTION OF STORMS AND THE AGENCIES WHICH APPEAR TO AFFECT IT	80
VII.	THE USE OF WEATHER CHARTS	102
VIII.	STORM WARNINGS	116

APPENDICES.

A.	WEATHER REPORT. NOVEMBER 29, 1874	148
B.	READINGS OF AUTOMATIC INSTRUMENTS. VALENCIA. MARCH 26, 27, 1874	151
C.	READINGS OF AUTOMATIC INSTRUMENTS. ABERDEEN. OCTOBER 20, 21, 1874	152
D.	READINGS OF AUTOMATIC INSTRUMENTS. FALMOUTH. FEBRUARY 1, 2, 1873	153

CHARTS.

Chart shewing the Position of the Telegraphic
Reporting Stations, January 1876 . . *to face Title*

Weather Charts, November 29, 1874, 8 A.M. (reduced
from the Daily Weather Report) . . *to face p.* 150

WEATHER CHARTS
AND STORM WARNINGS.

CHAPTER I.

THE MATERIALS AVAILABLE FOR WEATHER STUDY.

BEFORE proceeding to describe the charts themselves, or to give an account of their utility in aiding us to form a judgment as to probable weather, it will be well to explain the character of the observations and the nature of the service on which they are based, confining ourselves exclusively to the instruments employed and observations taken at our own telegraphic reporting stations (the outfit of which is necessarily far less complete than that of a fully furnished meteorological observatory); omitting technical descriptions of the instruments themselves.

The observations taken at these stations refer to Atmospherical Pressure, Temperature, Humidity, (or the Dampness of the air,) Wind, Rain, Weather, and, at sea-coast stations, Sea Disturbance.

The drawing of weather charts depends on the comparison of various observations taken at the same time, at several different stations. It is therefore indispensable, that all the observations to be compared should be freed, as far as possible, from incidental inaccuracies and discrepancies, by reducing the conditions at each place to common standards of comparison, and this is done by applying the several instrumental and other corrections.

Atmospherical pressure is measured by the barometer, and by this is meant the mercurial barometer; for the aneroid, however convenient it may be for ordinary use as a weather glass, cannot be recognised as an independent instrument, its indications depending entirely on the delicacy of the mechanical appliances employed in its construction.

The barometer has a thermometer attached to it, in order to show the temperature of the instrument itself, and this must be read whenever an observation of the height of the mercurial column is taken.

The readings of the barometer are always said to be 'corrected and reduced to 32° and to sea level,' and it is necessary to explain these expressions.

'Corrected,' means that corrections have been applied,

a, for the error of the scale of the instrument, which has been ascertained by sending it for verification to some recognised establishment, such as Kew Observatory,

b, for what is called 'capillarity,' which depends on the bore of the tube, and

c, for what is called 'capacity,' which depends on the

proportion which the bore of the tube bears to the sectional area of the cistern.

'Reduced to 32°,' means corrected according to the reading of the attached thermometer. The column of mercury in the barometer tube behaves like almost all other bodies, being lengthened by heat, and shortened by cold. It is therefore obvious that unless two similar barometers be precisely at the same temperature, they cannot possibly read alike, and hence is apparent the absolute necessity of reducing all readings to the same common temperature of melting ice, 32° F., which has been unanimously adopted as the best standard temperature for the purpose.

The temperature of the column being therefore indicated by the reading of the attached thermometer, according as that reading is above or below 32° F. we can discover, by means of printed tables, how much the column is too long, or too short, as compared with what it would be at 32°, and consequently what correction is required for the barometer reading.

'Reduced to sea level,' is a phrase which requires rather more explanation than the foregoing. The barometer, as its name implies, measures the weight of the air, and that weight, of course, depends mainly on the quantity of air which is above the instrument pressing upon the cistern, and keeping up the column of mercury in the tube. If, therefore, two similar barometers be placed one directly under the other, say, one in the attic, and the other on the ground floor, it is evident that there will be a less quantity of air above the former, than above the latter instrument, and consequently, the barometer in the attic will read lower than that on the ground floor.

Precisely the same reasoning will apply if we are considering two barometers at different heights on the side of a mountain, or one at an inland station, and the other at the level of the sea; in every instance the barometer at the higher station will read lower than that at the lower.

Hence we find that all readings must be reduced to their equivalent readings at a standard elevation, as well as at a standard temperature; this standard elevation is the mean level of the sea, and the reduction is carried out by means of tables.

A slight consideration of the foregoing remarks, and of the last-named principle in particular, will show how it comes about that the reading given in a Daily Weather Report for an inland station, like Oxford, or Nottingham, can never agree with the actual reading taken by an observer at either of those stations at the same time, unless the latter reading be corrected and reduced to 32°, and to sea level.

Temperature.—This does not require much explanation. In addition to the actual readings of the thermometers, 'in the shade' (*i.e.*, suspended in a properly constructed louvre-boarded screen at a height of four feet above the ground), which are taken at certain definite hours, the Daily Weather Reports give also two columns, showing respectively, the maximum and minimum readings taken during the twenty-four hours, ending at 8 o'clock each morning, by thermometers so constructed as to register, one the highest, and the other the lowest, temperature reached during the interval which has elapsed since the last setting.

The last-named readings may also be turned to very useful purpose. The mean temperature of the day is often mentioned in connection with sanitary statistics, as for instance in the Registrar-General's Reports. This mean temperature is theoretically the average of twenty-four thermometer readings taken at hourly intervals during the day. There are well-known rules for determining this mean temperature out of various combinations of hours of observation. The simplest method, and, for practical purposes, nearly the most exact, is to take the average of the maximum and minimum readings given in the Daily Weather Report, on any day, and put that down as the mean temperature of the preceding day.

Humidity.—The amount of moisture in the air is measured by the reading of a thermometer with its bulb encased in muslin, and kept damp (a wet bulb), compared with that of an ordinary (dry bulb) thermometer, without any such mounting, taken at the same moment.

This is not the place to enter into the theory of these observations. Suffice it to say, the greater the difference between the readings of the two thermometers the drier is the air, and when the two thermometers read alike, the atmosphere is exceedingly damp. The difference between the two thermometers ranges in this country from 0° to 10° or even 15°, and sometimes a difference of upwards of 20° has been noticed.

The chance of rain depends to a great extent on the degree of humidity of the air from time to time, and if we are dealing with reports from an extensive tract of country, as North America, or the Continent of Europe, the distribution of the moisture, or of the vapour tension will afford great assistance in tracing out the probable motion of storms. As regards these islands, the fact that most of the telegraphic stations are on the sea-coast detracts from the value of the reports of humidity, inasmuch as the amount of moisture in the air is seriously affected by proximity to the sea; and so we are unable to use reports of vapour tension in weather study as much as our neighbours.

Rain is measured by means of a rain-gauge. The values given represent the depth of water which would have accumulated on a level piece of ground, if none of the rain which fell could have escaped by drainage, etc. Too much faith must not be placed on the rain returns in the daily reports, for two reasons. Firstly, the gauges are often necessarily placed in towns, where a good exposure is not to be had, so that the amount measured is not quite the same as might be yielded by a gauge in a more open situation; and secondly, the stations are so sparsely distributed over the kingdom that it is impossible that they should give a precise account of the rain-fall in every county. What they do show is whether or not the rain is general and heavy.

Wind.—As a general rule, for the purposes of the Daily Weather Report, this is not measured by an instrument, but is simply estimated according to the Beaufort scale, which is so named after the late Admiral Sir Francis Beaufort. The following is the scale,[1] with the approximate equivalent velocity of the wind in miles per hour, as determined in the Meteorological Office.

				Miles per hour.
0.	Calm			3
1.	Light air	Or, just sufficient to give steerage way		8
2.	Light breeze	⎧ Or, that in which a well-conditioned man-of-war,	1 to 2 knots	13
3.	Gentle breeze	⎨ with all sail set, and 'clean full,' would go in	3 to 4 knots	18
4.	Moderate breeze	⎩ smooth water from	5 to 6 knots	23
5.	Fresh breeze	⎫	Royals, &c.	28
6.	Strong breeze	⎬	Single-reefed topsails and topgallant sails	34
7.	Moderate gale	⎨ Or, that to which she could just carry in chase, 'full and by'.	Double-reefed topsails, jib, &c.	40
8.	Fresh gale	⎬	Triple-reefed topsails, &c.	48
9.	Strong gale	⎭	Close-reefed topsails and courses	56
10.	Whole gale	Or, that with which she could scarcely bear close-reefed main-top-sail and reefed foresail		65
11.	Storm	Or, that which would reduce her to storm-stay-sails		75
12.	Hurricane	Or, that which no canvas could withstand		90

---

[1] Since Admiral Beaufort's time there has been a great change in the rig of merchant ships by the introduction of double topsail yards. It seems therefore advisable to add to Beaufort's scale the amount of sail which his ship would have been able to carry had she been rigged with double topsail yards, but under all other circumstances the same. The change would only affect forces 6 to 10.

    6. Topgallant sails.
    7. Topsails, jib, &c.
    8. Reefed upper topsails and courses.
    9. Lower topsails and courses.
   10. Lower main-top-sail and reefed foresail.

It is obvious that as this scale refers to the rate of sailing of, or to the amount of sail carried by, a ship, it is at first sight not well suited for use at land stations; but experience has shown that the reporters' estimates are never very far from the truth, when once they have gained some practice in observing. There are objections to supplying instruments for the measurement of wind to telegraphic stations, owing to the great difficulties which would be experienced in erecting them in suitable positions. Anemometrical indications taken in the middle of a town are almost worthless, as the buildings produce such eddies that the true movement of the air cannot be ascertained from the instrument.

As regards the numbers of the Beaufort scale, it is those from 6 upwards which present the most interest. 6 is the lowest number which is taken in the Meteorological Office to justify the issue of a warning to the coast, and 9 is the lowest figure which by the regulations of the Board of Trade can be pleaded by a captain as 'stress of weather,' in case of casualty to his craft. These velocities are not uniform, like that of an express train. The statement of 56 miles an hour, means 56 miles in the hour, but during that hour the wind may have been gusty, and at times have had a velocity of near 100 miles an hour, while at other times its hourly speed may have scarcely reached 30 or 40 miles.

It may be remarked that if we employed the pressure, instead of the velocity of the wind, to measure the violence of a gale, we should be able to obtain a record of individual gusts, and in the opinion of many persons, such as engineers, &c., such information would be more practically useful to the public, than statements as to velocity. The reason that pressure anemometers are not generally adopted is, that as yet it has not been found possible to reason with sufficient confidence as to the pressure of the wind on a structure, such as a factory chimney, from the indications shown by a pressure plate of, say, one foot square in area. Till the influence of the size of the pressure plate on its indications, in a wind of given strength, has been thoroughly determined, it seems premature to recommend the employment of such instruments.

Weather.—The various observations which are comprehended under this general term are those which do not admit of instrumental record, such as the fact of the occurrence of thunder, &c. These are reported according to the subjoined notation, which, like the wind scale, is due to Sir F. Beaufort.

b Blue sky, whether with clear or hazy atmosphere.
c Cloudy but detached opening clouds.
d Drizzling rain.
f Foggy.
g Dark gloomy weather.
h Hail.
l Lightning.
m Misty hazy atmosphere.
o Overcast, the whole sky being covered with an impervious cloud.
p Passing temporary showers.
q Squally.
r Rain, continued rain.
s Snow.
t Thunder.
u 'Ugly,' threatening appearance of the weather.
v 'Visibility,' whether the sky be cloudy or not.
w Dew.

The only thing to be remarked about the use of these

letters in the Daily Weather Reports is that, formerly when two or three were entered to the same place on the same day, it was implied that the observations had been taken in the order in which they were printed. Thus *b.c.p.r.*, used to mean that during the interval which had elapsed since the previous report, the sky was at first 'quite clear,' then 'detached clouds' came over it, 'passing showers' ensued, at last turning to persistent 'rain' which 'continued' at the time of sending off the telegram. This is the method in which the letters are used in Appendix A. This practice has now been discontinued, and the reports give information as to the state of the weather at 6 p.m. and 8 a.m., as will be seen at p. 16.

Sea Disturbance.—Lastly, the table contains a column for the Sea Disturbance, of which there are nine grades, depending on estimation, like the scale for wind force.

0. Dead calm.	5. Rather rough.
1. Very smooth.	6. Rough.
2. Smooth.	7. High.
3. Slight.	8. Very high.
4. Moderate.	9. Tremendous.

The height of the waves not unfrequently gives most valuable information as to the force of the wind in the offing, but we must not place too much dependence on this particular indication, for many of our most serious storms have come on without any previous disturbance of the sea on our coast. This is due to the fact that the sea disturbance is caused by the wind outside, and if that wind is not blowing in the direction of the coast, it will not impel waves towards that coast. It will presently appear how it comes to pass that a wind may advance to a coast without blowing towards it, but for the present it may be said, that the fact of a wind blowing in the direction of these islands, and driving a heavy sea on parts of our coasts, is no positive proof that the wind itself will ever be felt on our shores. In fact, some of the heaviest seas reported on our western coasts are proved to have been caused by Westerly gales blowing far out in the Atlantic, which never reached Europe at all.

There are yet other remarks to be made as to the sea disturbance column, which is, after all, a portion of our table often consulted anxiously by such landsmen and landswomen as may have to cross the Channel. In the first instance, the sea close in shore, at a place like Dover, is often much calmer than it is outside, as the harbour is sheltered from several winds, especially northerly ones. Secondly, the roughness of the sea depends in great measure on the direction in which the tide is running at the time of observation. 'Wind against tide' knocks up a sea at once, and frequently the delay of a few hours may make all the difference between the miseries of sea-sickness, and the enjoyment of perfect comfort, if the tide has changed in the interval.

These, then, are the materials with which we have to deal in drawing charts, discussing weather, and in studying warnings to the coast, and it is self-evident that they are far from complete.

It is obvious that the amount of information as to the appearance of the weather at each station which can be conveyed by the three or four letters which, at most, are given in the proper column, is quite insufficient to convey to our minds any clear idea of what an experienced person would have gathered as to the general character and prospects of the weather at that station ; and yet the general appearance of the weather was all that our fishermen and pilots had to guide them for generations, nay, for centuries, before the barometer was invented. Right well did they then, and do their successors now, know how to profit by such signs!

In the first place the code is necessarily so condensed that it cannot give information as to the form of clouds, whether these be the highest 'mare's tail' or the lowest rain cloud, or whether there be more than one stratum. The value of such particulars for the purposes of weather study need hardly be mentioned. The character of the clouds, their changes, and their amount, afford to the practised eye almost the most valuable information attainable as to the condition of the atmosphere above us. Lastly, the code does not tell us anything about the motion of the clouds, so often the truthful indication of coming wind, though we do endeavour to obtain, by means of words added to the telegrams, information on this motion, when there is anything remarkable in it.

In these and all such particulars, the Weather Reports, no matter how correct they may be, are only a poor substitute for actual personal observation. Any one trying to form a correct judgment of the look of the sky from these alone is like a physician trying to deal with a case without a chance of a personal interview with his patient. What can a resident in an inland town like London know of the appearance of the weather on the sea-coast on any day from any telegram, no matter how detailed?

The reports are therefore incomplete as to quality and quantity, but in many respects they are capable of material improvement, especially if it should ever be rendered possible to devote more money to the service.

As to quality, the instruments *should* be automatic, so that by looking at the record traced mechanically, we should be able to notice all that had passed since the last observation. The reporters *should* be experienced observers, with outdoor occupations, so as to be constantly on the look-out for any change of weather which may take place, and therefore they should be selected from such a class of men as signalmen on the coast. It is obvious that such a weather report as can be given by a clerk who simply runs out to look at the sky just before filling up his despatch, cannot be of as much value as that of a man who has been in the open air watching the weather for an hour or so, or even for the better part of the day. As to quantity also, the reports are not sufficiently complete, both as regards time and space.

As regards time, they are not nearly frequent enough. The Signal Office at Washington receives three reports every day from each of its stations; but, as is well known, that office is most liberally supplied with funds by Congress. Our own Meteorological Office, however, can only afford one at 8 a.m. from most of our stations, and at best we only get additional reports at 2 p.m. and at 6 p.m. from a few places. As regards Sunday mornings, our information does not reach us till next day. The reason of our not making a change in this respect is that, if we did receive the reports on Sundays we could not use the information to much purpose, for we could not warn the coasts of storms, the local telegraphic offices being all closed on Sundays for nearly the whole day. We have therefore one chance in seven always against us in trying to keep pace with the weather changes. On Sunday evenings, however, we have lately received reports, but only from a few stations.

Lastly, as to space, the area over which our own network and our international exchange extends, is far too limited for us to gain a general idea of the conditions which are prevailing all around us, and it is on these that our weather in great measure depends. This might, to a certain extent, be remedied by an extension of our continental communications, but, as I shall point out later on, the information obtainable from Europe is comparatively unimportant when contrasted with what we can supply to our continental neighbours. What we really want, to give us a better insight into our chances of weather from time to time, is unfortunately unattainable, and that is a system of reports from stations in the

Atlantic, say, at a distance of 600 miles from our coasts, for most of our storms advance on us from the Atlantic. It is a problem as yet unsolved, to moor a vessel in 1,000 fathoms water, and to connect her with the shore by a telegraphic cable. The experiment of a floating telegraphic station, which was tried in 1869 at the entrance of the Channel, in much shallower water than 1,000 fathoms, was not encouraging, as will be explained later (p. 129).

Nevertheless, in spite of all the geographical defects of our insular position, and the difficulties which are unavoidable in the early stages of any branch of knowledge, I shall hope to show that satisfactory progress has been made in the science of Weather Telegraphy in these islands. When so much has been accomplished in the short space of fifteen years, since the service was first organised by Admiral FitzRoy in 1861, it may be hoped that with perseverance, and continued efforts to improve our methods, we shall one day arrive at some clearer knowledge of the laws of our storms and our weather.

The following is a specimen of the Daily Weather Report, as issued in January 1876. A chart showing the stations will be found as the Frontispiece.

DAILY WEATHER REPORT, JANUARY 17, 1876.

STATIONS	† YESTERDAY EVENING					8 A.M. TO-DAY									PAST 24 HOURS			
	Barometer at 32° F. and Sea Level	Temp. in shade	Wind			Weather	Barometer				Shade Temp.			Wind			Force (0 to 12)	Weather	Sea (0 to 9)	Shade Temp.		Amount of Rainfall
			Direc.	Force (0 to 12)	Weather	Reading at 32° F. and Sea Level	Difference from 8 a.m. yesterday	Dry	Wet	Difference from 8 a.m. yesterday	Direc.	Force (0 to 12)				Max.	Min.	
Haparanda	29·11	33	SW.	2	r	28·99	?	30·30	?	SW.	4	b	*	*	*	*	0·59
Hernösand	29·41	39	SW.	2	b	29·40	—·34	34·31	—2	SW.	2	b	*	*	*	...
Stockholm	29·71	36	SE.	6	o	29·67	—·32	33·32	0	WSW.	2	b	*	*	*	...
Wisby	30·04	34	WSW.	8	o	29·90	—·30	34·33	+3	WSW.	6	bc	*	*	*	...
Christiansund	29·44	40	WSW.	9	r	29·65	+·02	36·32	+10	W.	9	o	8	46	?	0·08
Skudesnaes	29·98	41	WNW.	6	b	30·00	+·01	39·39	—1	W.	4	bcf	4	41	37	...
Oxö (Christiansand)	30·01	41	WNW.	6	b	29·93	—·21	40·37	+1	WSW.	4	b	2	41	36	...
Skagen (The Scaw)	30·03	38	W.	4	b	29·95	—·25	39·37	+2	WNW.	6	b	6	39	32	...
Fanö	30·22	35	WSW.	4	bc	30·22	—·20	34·34	0	WSW.	4	m	*	36	32	...
Cuxhaven	30·35	*	*	*	*	30·27	—·24	43·43	+13	WSW.	2	f	*	?	25	0·13
Sumburgh Head	29·87	40	WNW.	6	b	29·98	+·24	41·40	+1	WNW.	1	bc	4	45	36	0·02
Stornoway	30·15	40	W.	3	b	30·03	—·05	40·39	—4	SW.	4	o	4	43	35	0·12
Thurso	30·07	41	WNW.	5	m	30·06	+·10	38·37	+1	SSW.	2	m	3	46	37	...
Wick	30·08	39	WNW.	5	b	30·07	+·05	38·36	—2	SW.	2	o	1	44	37	...
Nairn	30·18	41	SW.	2	b	30·14	+·03	39·35	—1	SW.	1	b	1	43	37	...
Aberdeen	30·19	42	WSW.	5	b	30·15	+·04	37·35	—3	WSW.	2	b	1	47	36	...
Leith	30·29	42	W.	1	b	30·24	...	41·39	...	WSW.	1	cm	*	49	38	...
Shields	30·34	40	W.	3	b	30·28	—·05	37·36	—4	SW.	2	bm	2	44	37	...
York	30·38	38	WNW.	1	b	30·31	?	36·36	+4	S.	1	f	*	40	31	...
Scarborough	30·35	38	WSW.	2	b	?30·39	+·03?	37·37	+4	WSW.	2	b	2	39	33	...

The Materials available for Weather Study.

Nottingham	30·46	35	SSW.	1	o	30·38	−·16	35/35	+10	WNW.	1	f	*	36	25	··
Ardrossan	30·32	46	WNW.	3	bc	30·23	−·05	44/42	0	SSW.	3	c	3	48	42	0·02
Greencastle	30·34	42	W.	2	b	30·21	−·11	38 37	−4	ENE.	3	o	1	51	33	0·02
Donaghadee	30·36	38	SW.	3	bc	30·26	−·11	37 36	−1	SW.	2	b	1	42	35	··
Kingstown	30·42	44	WNW.	3	b	30·30	−·14	42 40	−2	S.	1	cf	1	47	40	0·04
Holyhead	30·41	45	SW.	2	m	30·30	−·16	43 43	0	SW.	2	f	2	46	41	··
Liverpool (Bidston)	30·41	41	WSW.	3	o	30·32	−·16	40 38	+5	WSW.	3	c	4	43	34	0·21
Valencia	30·46	48	SSW.	3	m	30·31	−·20	50 50	+2	WSW.	4	f	4	51	34	0·07
Roche's Point	30·49	47	WSW.	4	b	30·32	−·24	49 49	+5	WSW.	3	o	3	49	44	0·07
Pembroke	30·48	46	W.	3	o	30·35	−·20	48 48	+7	WNW.	4	f	2	48	41	0·10
Portishead	30·48	40	SW.	3	o	30·36	−·24	43 43	+16	SSW.	2	m	2	45	26	0·02
Scilly	30·56	45	W.	3	o	30·45	−·19	49 48	+10	WSW.	4	o	4	49	47	0·03
Plymouth	30·55	35	Z.	0	o	30·43	−·20	39 39	+10	Z.	0	ofd	*	39	29	0·04
Hurst Castle	30·50	36	W.	4	o	30·40	−·21	41 41	+13	WSW.	2	o	1	41	25	··
Dover	30·49	36	W.	1	o	30·39	−·19	37 37	+9	SW.	1	m	*	40	26	··
London	30·50	31	W.	2	o	30·38	−·21	38 38	+13	Z.	0	o	*	38	25	0·02
Oxford	*	*	*	*	*	30·37	−·21	38 38	+16	SW.	1	f	*	38	22	··
Cambridge	30·45	30	SW.	1	o	30·37	··	35 35	··	W.	3	f	3	37	22	0·01
Yarmouth	30·41	30	W.	4	m	30·36	−·18	33 33	+7	WSW.	*	m f	2	34	25	*
The Helder	*	*	*	*	*	30·31	··	36 *	··	WSW.	3	*	2	*	*	··
Cape Gris Nez	30·51	32	W.	2	m	30·40	−·18	39 35	+12	WNW.	2	f	1	37	27	··
Brest	?	43	NNE.	1	bc	?	?	45 43	+11	Z.	2	f	2	?	?	*
L'Orient	?	34	E.	2	b	?	?	36 36	+6	ENE.	0	o	2	?	?	*
Rochefort	30·52	37	NNE.	4	o	30·56	+·07	25 ?	−5	NW.	3	o	1	39	28	*
Biarritz	30·49	43	ENE.	3	*	30·56	+·16	41 37	+7	··	3	o	2	43	30	?
Corunna	*	*	*	*	*	··	··	* *	··	··	··	··	··	*	*	*
Brussels	*	*	*	*	*	30·38	−·20	30 30	+5	W.	3	o	3	*	*	*
Charleville	*	*	*	*	*	30·40	−·11?	21 ?	+4	W.	1	r	*	*	*	*
Paris	30·56	21	NW.	3	s	30·47	−·09	28 28	+5	SW.	2	f	*	*	*	?
Lyons	*	*	*	*	*	?	?	33 *	··	S.	2	o	3	*	*	*
Toulon	30·20	46	N.	2	b	30·20	+·05	41 39?	−4	NW. W.	2	b	3	50	?	··

Weather Charts and Storm Warnings.

Daily Weather Report, Jan. 17, 1876—continued.

JAN. 16, 2 P.M. REPORTS AND REMARKS.

Stations	Bar.	Dry	Wet	Wind		Wea.	Sea
Skudesnaes . .	29·96	41	39	W.	6	b c f	6
Thurso . .	...	...	...	...	...	...	...
Scarborough .	...	...	...	...	...	...	...
Greencastle . .	...	...	...	...	...	...	*
Holyhead . .	...	...	...	...	...	...	...
Valencia . .	...	...	...	...	...	...	...
Scilly . . .	...	...	...	...	...	...	...
London . .	...	...	...	...	...	...	*
Rochefort . .	30·46	32	32	NE.	5	b c	3

EXPLANATION OF COLUMNS.

WEATHER.—*Beaufort scale* is : b, blue sky ; c, detached clouds ; d, drizzling rain ; f, fog ; g, dark, gloomy ; h, hail ; l, lightning ; m, misty (hazy) ; o, overcast ; p, passing showers ; q, squally ; r, rain ; s, snow ; t, thunder ; u, ugly, threatening ; v, visibility, unusual transparency ; w, dew.

* *An Asterisk* is inserted in all places for which information is not usually received.

† The evening observations are taken at 6 p.m. in our Islands, at 8 p.m. (Christiania time) in Sweden, Norway, and Denmark, and at 7 p.m. (Paris time) in France.

The report being for a Monday, the 2 p.m. reports of the previous day, Sunday, are defective.

It will be seen that in addition to the absent observations, which are marked with an asterisk, as is explained at foot of the table, various spaces are marked with dots (...), indicating that the report had not arrived for that day ; and others with notes of interrogation, indicating that for some reason or other the report appeared to be doubtful, in which case it is sometimes inserted and queried, and at others omitted altogether.

CHAPTER II.

THE WIND.

HAVING thus discussed the observations themselves, I come to the results and conclusions we can draw from them, and in order to make these intelligible, I must indulge in a little theorising, and refer to some things which, though not yet recognised as absolutely proved, appear at least probable, as regards the behaviour of the wind.

Every one knows that the East is very different in its character from the West wind, the former being reputed to be 'good neither for man nor beast.' Any one with a touch of bronchitis or rheumatism can (or at least *thinks* he can) tell you, without looking at a weather-cock, whether or not there is easting in the wind. This contrast arises from the fact that as a rule the temperature and dampness of the air in Western Europe are both lowest when the wind is about North-east, and highest when it is about South-west. This is, however, a local phenomenon peculiar to certain parts of the globe, for if we travel from Western Europe, on the same parallel of latitude as our own, either eastwards to the Sea of Ochotsk or westwards to Labrador, we shall find that in these districts the coldest wind is near North-west, and the warmest about South-east. In each case the coldest and driest point of the compass lies towards that region in the neighbourhood of the point of observation where the mean temperature is the lowest. In these islands, in winter, this region is northern Russia, in Labrador it is the Barren Grounds of the Hudson's Bay Territory, and at the mouth of the Amur it is the district of Yakutsk. These two latter cold regions lie to the north-west of the respective coast districts to which reference has been made.

The wind, then, is cold and dry when it comes from a cold region, warm and moist when it comes from a warm district, such as the sea surface in these latitudes in winter. In summer there is not so strong a contrast between the temperature of different parts of the earth's surface in the northern hemisphere as in winter, and so the different winds do not differ so much in their characters.

For many years it has been the fashion to say, that all cold winds flowed from the Poles to the Equator, forming the so-called Polar Currents, and becoming the Trade Winds when they approached the Tropics, while the warm winds flowed from the Equator to the Pole, forming the Equatorial Currents, or Anti-trades.

There can be no doubt that these statements are right in principle, as the original disturbing action which produces motion in the atmosphere is the heat of the sun; and this acts most strongly in the Torrid zone, which ought, accordingly, to be the region towards which all the cold air flowed, were it not for the irregular arrangement of land and water on the globe. Recent investigations have, however, shown that a body of air can hardly ever be proved to have made its way direct from the equator to the pole, inasmuch as it will probably have been caught up on its way by some of the eddies and local circulations at all times existing on the earth's surface. The motions of the atmosphere are found to be mainly regulated by the distribution of barometrical pressure over the globe, the particles moving from the regions where the pressure is high to those where it is low, and being modified in the direction of their motion by various causes, among which the earth's form, and its rotation on its axis, are the most influential.

I have said that a body of air can hardly be proved to flow the whole way from the equator to the poles or back again, but nevertheless it is the fact that over extensive areas of the earth's surface the wind does maintain a constant direction for a considerable period of time. To prove this it is only necessary to cite the well-known phenomena of the Trade Winds and the Monsoons, where the wind blows persistently from the same direction, in one case for the whole year, and in the other for months at a time. In our own latitudes the phenomena of the winds are not so regular, but yet it is found that over large tracts of Europe and the North Atlantic, the wind at times blows for weeks together in a definite direction, either Easterly or Westerly, (speaking in general terms,) and that these tracts, or the channels of these great currents, lie the one alongside the other. The disturbances which cause our storms appear to occur along the debateable regions between two such currents, and when the currents change their beds, or the lines of demarcation between them alter their positions, the storms and disturbances move with them.

It will be explained subsequently how the motion of the air, both in direction and velocity, is regulated by the distribution of atmospherical pressure at the surface of the earth, which is shown by the distribution of the readings of the barometer in the weather chart; but as this relation lies at the very foundation of the whole structure of modern weather knowledge, it will be useful, at the very outset, to state it, even at the risk of having to repeat my words more than once in the succeeding pages.

This principle, which, for convenience, is known as Buys Ballot's Law, is contained in the following statement.

Stand with your back to the wind, and the barometer will be lower on your left hand than on your right.

These words hold good, except close to the equator, for the northern hemisphere; in the southern we must interchange left and right. If this principle be once thoroughly recognised, the broad features of wind motion will be at once understood.

The *Force* of the wind, as distinguished from its *Direction*, is related to the amount of difference of barometrical pressure over a given distance, and this is defined as the 'gradient,' a term which will be explained in Chapter IV. p. 41, so that this force in no way depends on the absolute height of the barometer at any one station, which the words printed on the scales of many old-fashioned barometers would seem to indicate. In fact, as an illustration of this statement, I may anticipate my future explanation by stating that at Liverpool, during a severe storm on January 24, 1876, from 2 till 3 a.m., the velocity of the wind from SW. was sixty-two miles within the hour, while the barometer was 30·10 inches, nearly at 'Set Fair.' Conversely, on a still more recent occasion, at 6 p.m. on March 9, 1876, the barometer at Wick read 27·94 inches, far below 'Stormy,' and the force of the wind was only 3, (a 'gentle breeze,') from NW. In the one case, therefore, we had a heavy gale with a high barometer; in the other, a gentle breeze with a very low barometer. The reason of these discrepancies between old theories and actual facts will appear when we treat of gradients.

Storms were formerly divided into two great classes, circular storms (hurricanes and typhoons), and straight-line storms. The former are almost the only class of storms which occur within the tropics, and are known under the general name of cyclones. The latter class were formerly supposed to be the most usual type of storms in these latitudes, inasmuch as it is a common observation here that the wind will blow hard from the same point, and for a considerable length of time, over a large district. The study of observations taken at the same hour over an extensive tract of the earth's surface has, however, shown that the storms of the Temperate zone are almost without exception cyclonic, or partially so, in their character, although not so perfectly developed as those within the tropics. All cases of so-called straight-line storms are to be explained either by the persistence of the same characteristics for several days over the same region, or else simply by the fact that they are mere local phenomena, due to the contour of the country, like the exceptionally strong breezes often met with on rounding prominent bluff headlands, or, to use a more familiar illustration, at street corners. By this statement it is not meant to imply that strong winds from a definite point are not met with for days together in the region of the Trade Winds and Monsoons; but, firstly, these forces never reach those of an actual 'strong gale,' and, secondly, the conditions which cause them are of the same nature as those which cause our own storms, and it is the persistence of the conditions which determines the constancy in the force of the wind.

In speaking of the general character of storms, it should be mentioned that in most cases the vertical depth of the stratum of the atmosphere which is in the condition of storm is very small in comparison with the superficial extent of the area over which the storm is felt. Everyone knows that it is a common occurrence to see clouds at a moderate elevation either moving rapidly while calm prevails below, or else at rest or nearly so, while we are feeling a strong wind. On a recent occasion, August 18, 1875, when a tornado passed over a village in Sweden called Hallsberg, in the province of Nerike, it was expressly noticed that while the branches of trees and fragments of the wreck of buildings were carried by the wind for miles, the clouds did not indicate the slightest sign of disturbance.

I have already spoken of great currents of air extending over vast tracts of the earth's surface, and the best idea which we can gain, for practical purposes, of the winds which affect us in these islands, is that the air over the Atlantic Ocean, north of latitude 40° north, is constantly flowing from west to east, like a gigantic river. If such a river be flowing rapidly, we often see on its surface small waves, each with its own eddies and circulations, which are carried on with the stream. If we could look at the upper surface of the atmosphere, we should see much the same sort of conditions, except that what corresponds to the hollow of the wave would be a patch of defective pressure, while that which corresponds to the crest of the wave would be an area of excessive pressure. We shall shortly learn how these areas influence the motion of the air and the weather.

There remains one general principle which has been often brought forward as almost indisputable, and that is the principle known under the name of the 'Law of Gyration,' propounded by Professor Dove, of Berlin, some forty years ago. This principle is, that the wind changes more frequently 'with the sun,' that is, from east, *through south*, to west, in the Northern Hemisphere, than it does in the other direction, and that this is true all round the compass. The enunciation of this doctrine as an universal law has arisen from the fact of meteorology having been first studied in Western Europe, where the truth of the principle is undeniable. When, however, we look to the evidence cited by Professor Dove himself for Arctic stations, we find that the statements from those regions are not nearly so positive in favour of the law as the experience gained in lower latitudes, and the evidence from the German Arctic Expedition shows that on the east coast of Greenland the direction of change of wind was more frequently *against* the sun than *with* it.

Moreover, when we come to consider the motion of wind in the systems of high and low pressure, of which I have just been speaking, and the motion of these systems themselves over the earth's surface, we shall find that the question of the shifting of the wind in accordance with the law ('veering'), or its shifting in opposition to the law ('backing'), simply depends on the motion of the systems of circulation to which the winds in question belong. For instance, a wind 'veers,' or shifts, 'with the sun,' at any station when an area of low pressure passes from west to east to the northward of that station, and it 'backs' when the area of low pressure passes in the same direction to the southward of the station. See fig. 14, p. 73.

The reason that veering, especially from SE. or S. through SW. to W. or NW., is so common in these islands, is that the most usual track of cyclonic systems of air is from SW. to NE., the centres of the storms passing to the northward of most stations in this country.

We have learnt, therefore, that wind is always connected with some disturbance of the pressure of the atmosphere, and it will be at once understood that its existence is due to the tendency of an elastic fluid, like the air, to regain the condition of equilibrium from whence it has by any means been disturbed, while its motion is regulated by certain fixed laws to which we have alluded, and which will be more fully stated presently.

CHAPTER III.

THE BAROMETER.

WE have to consider next what the nature of these disturbances of which we have been speaking really is. They are connected with irregularities in the distribution of atmospherical pressure. I only say 'connected with,' and do not say 'traceable to,' or use any more positive expression of opinion, for I do not propose to discuss the question of what the original causes of the disturbances are. Various theories have been propounded to account for storms, and some will be mentioned in Chapter VII., but none of them have met with general acceptance as yet. We must therefore only take things as we find them, and endeavour to make the best of them.

But it is necessary, at the outset, to explain the meaning of some of the terms which will recur most frequently in the following pages.

The first of these is the word 'isobar,' which is derived from two Greek words signifying 'equal weight.' *An isobar, or isobaric line, is a line passing through those places where the barometrical pressure is equal.*

If we look at the chart (fig. 1, p. 30) containing the readings from the Report given on pp. 16, 17, we see that the readings in question, all of which are shown on the chart, varied from 29·65 inches at Christiansund in Norway, to 30·56 at Rochefort and Biarritz in France. We find several lines drawn across the chart: all these lines are isobars, and the values of each are given.

The courses of these isobars are determined with reference to the stations on each side of them where actual observations exist. Thus the isobar of 30·1 passes in Scotland between Nairn (30·14) and Wick (30·07), and it reaches the Danish coast between Fanö (30·22) and the Scaw (29·95), its probable position in each case being ascertained by dividing the distance between the stations in question in proportion to the difference in readings between them. Similarly, points are found for the other isobars crossing the North Sea by dividing the distance between the stations on each side of that sea in proportion to the difference in readings between them. In this manner the courses of isobars drawn over the sea, and in regions whence no observations are obtainable, are inferred from the readings taken at the nearest stations whence observations exist. Thus it is quite easy to trace the course of the isobar of 30·4, where it is prolonged outside the coast of Cornwall. As a general rule isobars drawn on comparatively insufficient information are dotted instead of being drawn in full.

The isobar of 30·5 inches takes a sweep over the western parts of France, but is not continued over the

30   *Weather Charts and Storm Warnings.*

sea, or over Spain, from want of information. That of 30·4 passes between Plymouth (30·43) and Portishead (30·36), through Cape Gris Nez and Charleville (both

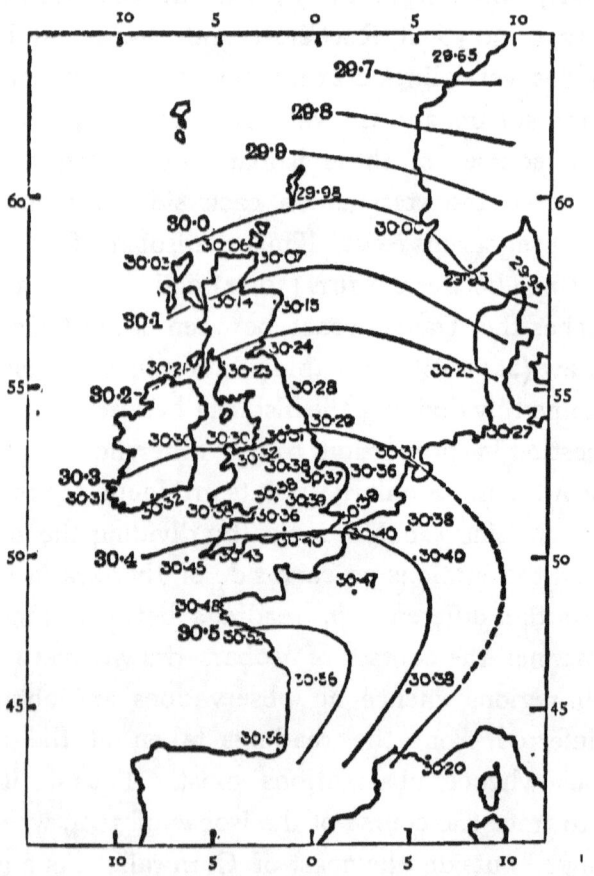

FIG. 1.—January 17, 1876, 8 a.m.   Barometrical Readings and Isobars.

30·40), and ends nearly half way between Toulon (30·20) and Biarritz (30·56). The line of 30·3 is very clearly marked; it starts close to Valencia (30·31), through

Kingstown and Holyhead, close, on the northern side, to York, Scarborough, and the Helder (all three 30·31). It is then dotted, from deficiency of information, and carried out to the neighbourhood of Marseilles. That of 30·2 passes through Glasgow and St. Andrews to Denmark, a little north of Fanö (30·22). That of 30·1 crosses Ross-shire and the Moray Firth to the west point of Jutland, while the line of 30 inches sweeps just south of Sumburgh Head (29·98), through Skudesnaes and south of the Scaw (29·95). There are three other lines indicated for a short portion of their course over Norway, inasmuch as the nearest reading to Skudesnaes on the northern side is Christiansund (29·65), and so the difference of three and a half tenths of an inch between these stations necessitates the drawing of three lines.

From what has been said, it will be understood that as these lines pass between stations where the observations show higher or lower pressures respectively than the values assigned to the isobars themselves, it is presumable that the reading at any other station exactly on any line would have precisely the value shown by the isobar.

These isobaric lines are drawn on all weather charts, and a knowledge of their respective values in barometrical readings, and of their courses, lies at the very foundation of all that we know about weather.

It will already have been seen from fig. 1, that the readings are not the same at all stations simultaneously, and if we look at any chart representing the conditions of barometrical pressure over an extensive region for any epoch of time, we shall find that this pressure is very far from being equal at all stations, so that it is greater in some places than in others. If we then proceed to draw the isobars, we shall find that several of these lines form closed curves round certain spots where the barometrical readings are either *lower* or *higher* than they are over the neighbouring districts.

These two contrasting states of affairs are known by different names. The districts of low pressure, or of depression, are termed 'cyclonic,' and those of high pressure 'anticyclonic,' the names being derived from the Greek word 'cyclos,' a circle, and expressing the fact that the wind in both classes of areas has a tendency to circulate round the centre of the system of disturbance, but in opposite directions in the two cases.

There is a very marked difference between the areas in which the barometer is lower, and those in which it is higher, than in the surrounding districts, for the temperature and the weather, as well as the circulation of the wind, differ most materially under the respective conditions.

In both cases there is a calm at the centre, or over the region enclosed by the innermost isobar, and, as a general rule, in the cyclonic systems the extent of this calm centre is less, and the isobars surrounding it lie closer together than in the anticyclones. We shall shortly learn that this indicates a material difference as regards the force of the wind.

In order to do this it will be necessary to give some charts as instances of cyclonic and anticyclonic disturbances respectively, which have existed over the region embraced by our telegraphic reporting system. It is, however, comparatively seldom that the whole of a system will be found developed over our limited district. More frequently we only find curves trending one way or the other, which enable us to conclude, from the general distribution of pressure, in what direction the central area of high or of low pressure, as the case may be, is situated.

Fig. 2 (p. 34) gives a good example of an area of low pressure, or a 'depression,' or a cyclonic disturbance, for the terms are used almost indiscriminately, developed as fully as it is usual to find them in Western Europe. Since the mode of drawing isobars has already been explained, it has not been thought necessary to reproduce the actual barometrical readings at the different stations.

It will be noticed that the chart contains other indications besides those afforded by the isobars which require some explanation.

The direction and force of the wind are given by arrows, or by a circle if there is a calm.

The direction is of course shown by the direction in which the arrow is flying.

The force is indicated by differences in the symbols employed, which are as follows :—

	Forces 0-1 (Beaufort scale, p. 7)	⊙
,,	2-4 ,,	→
,,	5-7 ,,	→
,,	8-10 ,,	↦
,,	above 10 ,,	↦

D

Thus it will be seen that there is a very heavy gale at Rochefort from WNW., a heavy gale at Scarborough from SE., a fresh breeze at Aberdeen from E., a light

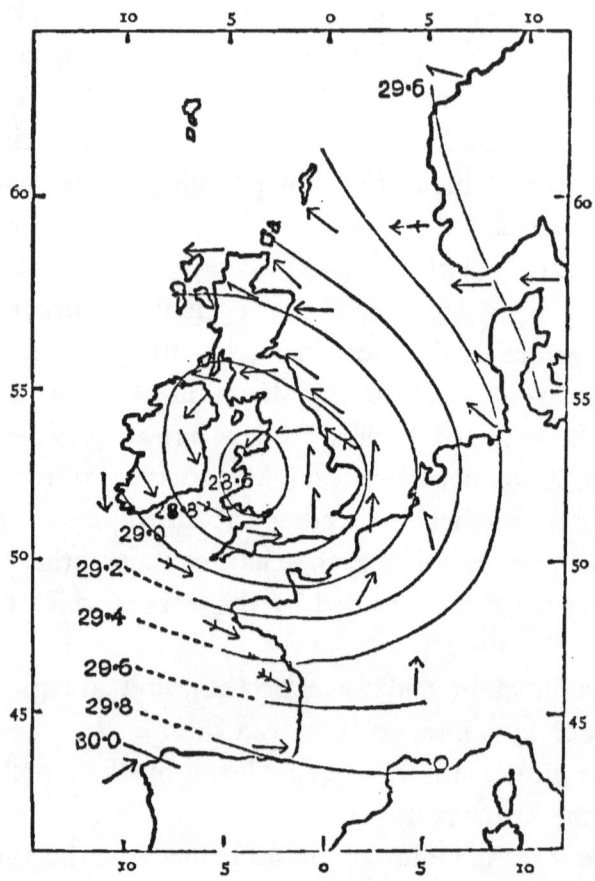

FIG. 2.—November 29, 1874, 8 a.m.   Isobars and Wind, Cyclonic System.

breeze at Brussels from SSE., and a calm at Toulon. (See Appendix A.)

The Barometer.

The lowest reading (28·55 inches) is at Holyhead; the highest (30·00 inches) at Corunna.

The innermost isobar (28·6) embraces almost the whole of Wales. That for 28·8 is oval in shape, and covers nearly all England, and the east and north of Ireland. That for 29·0 takes in a little of France and Belgium, and the greater part of Scotland. The isobar of 29·2 envelops the whole of Scotland, but is not carried out over the Atlantic beyond the Orkneys on one side and the coast of Brittany on the other, and it is only dotted in, as being merely inferred, in the absence of observations over the Bay of Biscay.

The other isobars are only partly shown; in fact the position of a portion of that for 30 inches is only indicated by the single observation of 30·00 at the station of Corunna.

The chief feature noticeable in the northern part of the chart is that the isobars of 29·4 and 29·6 trend away to the northward, and so the curves spread out in a fan shape between the Shetlands and the coast of Norway.

If we now turn to the wind arrows, we shall find that they show a circulation round the centre of depression. They are:

Westerly at Scilly and in France,	on its southern side.
Northerly at Valencia, North-north-east at Donaghadee,	on its western side.
East at Ardrossan,	on its northern side.
South-east along the whole east coast of Great Britain north of Hull, South at Yarmouth and in the Straits of Dover,	on its eastern side.

In fact the wind sweeps round the central area of depression, *against watch hands*, and this is the invariable law in all cases of cyclonic disturbances in the northern hemisphere. The wind moves in a direction opposite to that of the hands of a watch, and its course is nearly parallel to the isobars; but to this subject, as well as to the relation of the force of the wind to the distribution of pressure, we shall return later on.

Let us now take the converse case, an instance of an area of high pressure, or an anticyclone, and for this we have an excellent example in February 4, 1874 (fig. 3). On this day at 8 a.m. the absolute highest reading is 30·67 at Nottingham, and the only English reports which give readings below 30·6 are Dover, Plymouth, and Scilly, the isobar of 30·6 enveloping almost the whole of England and Wales. That of 30·5 stretches to Holstein near Fanö (30·49), passes south of Paris (30·52), being shown by dots owing to deficient information between these points, sweeps close to Valencia (30·49), is again dotted over the sea outside the coast of Ireland, and finally reaches Aberdeen (30·49).

In the north and south readings decrease rapidly; on the south side 30·40 passes half way between Rochefort and Biarritz, and 30·30 skirts the coast of Spain and the Pyrenees, readings being 30·28 at Corunna, and 30·24 at Toulon. In the north 30·4 passes below Stornoway (30·39), runs between Wick and Thurso and across to Jutland, where the reading at the Scaw is 30·34.

Passing further north, 30·30 runs close to Sumburgh Head and across to the neighbourhood of Bergen in Norway, while above it still we find that of 30·20, the reading at Christiansund being 30·15.

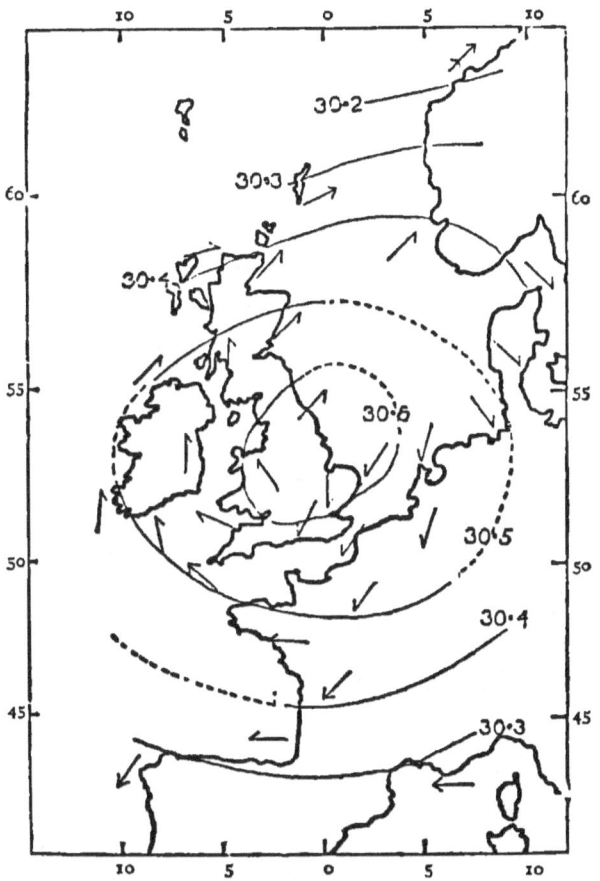

FIG. 3.—February 4, 1874, 8 a.m. Isobars and Wind, Anticyclonic System.

When we look at the winds in this case we find that

their circulation is exactly opposite to that which is shown in fig. 2 ; it is *with watch hands*, being :

North	in Germany, on its eastern side.
East	in France, on its southern side.
South	in Ireland, on its western side.
South-west to west	in Scotland, on its northern side.
North-west	in Denmark, on its north-eastern side.

The arrows also appear to draw *out from the centre* instead of *drawing in towards it* as in cyclonic systems.

The force is far lighter than it was in the former case, a characteristic of anticyclones as compared with cyclones. This, as will shortly be explained, is owing to the fact of the isobars being far apart, a marked feature of anticyclonic systems.

We have therefore learnt in this chapter (1), that the distribution of atmospherical pressure on a weather map is made clear by the isobars; (2), that there are two grand types of disturbance of the equilibrium of pressure, called respectively, *a*, cyclonic and, *b*, anticyclonic, according as the pressure in a given locality is either in defect or excess of its mean value in the surrounding region; (3), that these two types are characterised by strong contrasts to each other, among which the most striking are the differences in the direction of motion of the wind.

Cyclonic and anticyclonic are, however, merely *relative* terms ; the reading may be above 30 inches at the centre of a cyclonic area, and below 29·5 inches at the centre of an anticyclonic region. If at any place, or over any district, the barometrical reading is *lower* than at the places all round it, that place or district is the centre of a cyclonic area : if, on the contrary, the reading at the place or district is *higher* than at the places all round it, that place or district is the centre of an anticyclonic area.

CHAPTER IV.

GRADIENTS.

LET us now proceed to examine still more closely into the principles in regard of the motion of the wind, which have only been faintly hinted at before. The direction of motion of the wind, in relation to the distribution of atmospherical pressure, may be easily perceived to be such that if you stand with your back to the wind the barometer will be lower on your left hand than on your right. This is the principle already cited, and generally known, as Buys Ballot's Law, from the name of the Dutch meteorologist who first insisted on its universal applicability. It is simply the extension, to all cases of wind motion, of the Law of Storms first announced by Redfield and Reid for the Hurricanes of the West Indies and of the Southern Indian Ocean, as well as for the Typhoons of the China seas. The direction of rotation is opposite in the two hemispheres. The law as above stated refers to the Northern Hemisphere. The law therefore gives the *direction* of the wind, but it also enables us to judge of its *force*, for it is found that this latter depends mainly on the amount of difference of pressure between adjacent stations.

Gradients.

Air, being a gas, is even more mobile than water, and as the least difference in level between two portions of a free surface of the latter generates motion, tending to produce equality of level; so in the former case the slightest difference of pressure causes motion in the atmosphere, in the endeavour to restore equilibrium of pressure.

It is evident that the greater the difference of pressure over a given distance, the greater will be the effort, and consequently the more violent and rapid will be the motion, required to regain equilibrium. There is therefore a convenience in fixing a standard of comparison by which to measure the disturbance of pressure, and here meteorologists borrow an idea from engineers, who measure the inclination of a road or railroad by what is termed the 'gradient,' implying, when they speak of a gradient of one in sixty, that the slope rises one foot vertically for every sixty feet of horizontal measurement.

Meteorologists speak of gradients also, but instead of applying the same unit of measurement, as of feet, to the vertical and horizontal scales, they give the vertical scale in units of barometrical measurement, and the horizontal scale in miles of distance. These gradients therefore are expressed in differences of barometrical pressure over a given distance.

The gradients adopted by the Meteorological Office are expressed in hundredths of an inch of mercury per one degree of sixty nautical miles.[1]

---

[1] On the Continent gradients are measured as millimetres per one

In the accompanying figure (fig. 4), the horizontal distance between the two stations A and B is supposed to

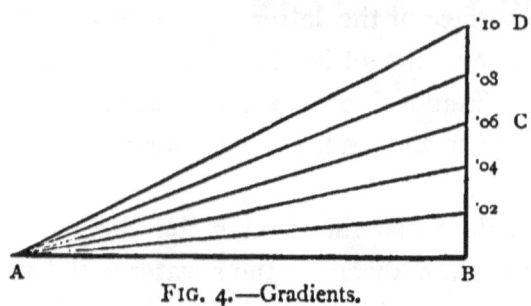

FIG. 4.—Gradients.

be sixty nautical miles. The divisions on the vertical line B D are hundredths of an inch, and they correspond to the differences between the barometrical readings taken at the same hour at the two stations. The gradients are the ratios between the intercepts B C, B D, &c., and the line A B which is supposed to be = 1. The gradients are given as 6 for the angle B A C, 10 for the angle B A D, corresponding to the several observed differences. These lines A C, A D, &c., are imagined to be drawn every morning between the most important stations given in the Daily Weather Report, and from their inclinations conclusions as to the probable direction and force of the wind for the day are drawn. It is found, for instance, that the force of the wind will not exceed the figure 5 or 6, a 'fresh breeze' on Beaufort's scale,

degree of sixty nautical miles. These gradients are therefore about one-fourth the magnitude of ours.

The unit of distance for gradients formerly used in this country was fifty nautical miles; the change to sixty has been made for greater uniformity with foreign nations.

Gradients. 43

unless the gradient be higher than 6 (A C on the diagram).

To reduce this statement to a practical form, I may put it in these words. The distance from Penzance to Brest is 113 nautical miles. A gradient of 7 between these stations represents a total difference in barometrical readings of 0·13 inch, so that, in accordance with what has just been said, whenever a Westerly gale is blowing at the entrance of the Channel we may expect that the barometer at Penzance will be at least 0·13 inch lower than that at Brest; *vice versâ*, the readings at Brest will be proportionably lower than those at Penzance whenever an Easterly gale is felt in the district in question.

An instance in point for the former state of affairs is the gale of January 8, 1870, during which the reading at Brest at 8 a.m. was 29·38, and that at Penzance, at the same hour, was 29·19. The difference between these readings is 0·19, and the resulting gradient 10. A very severe Westerly storm was that morning reported from the Channel.

The converse conditions, accompanying an Easterly gale, were observed on May 14, 1869, when the reading at Penzance was 29·92, and that at Brest 29·68. The resulting gradient is nearly 13, and accordingly heavy Easterly gales were felt on our Channel coast.

To apply the same principle to the winds of the British Islands generally, it may safely be asserted that no storm of any serious extent is ever felt over the

United Kingdom unless there be an absolute difference in barometrical readings exceeding half an inch of mercury between two of our stations.[1]

The difference in readings between Rochefort and Aberdeen on February 1, 1868, when a tremendous Westerly gale was raging, was as much as 1·76 inch: the reading at Rochefort being 30·16, and that at Aberdeen 28·40 inches. These figures give a gradient of 15·7 over the entire distance of 673 miles, and we find that gales were reported from seventeen stations that morning.

No very precise relation has as yet been established between the amount of the gradient and the force of the wind, if such exists, but as a convenient figure to be remembered I may repeat that a gradient of 0·07 inch per 60 miles indicates the probability of as much wind as an ordinary yachtsman likes to meet with.

We are now in a position to see more clearly how entirely this idea of gradients does away with the old notion that the actual height of the barometer at one station gives a certain indication of the probable direction or force of the wind or of the character of the weather at that station, a notion which has found expression in the words 'Very Dry,' 'Fair,' 'Change,' 'Rain,' 'Stormy,' &c., &c., which have been for so many years placed on barometer scales.

[1] Local storms, which occasionally do great damage, may be felt when the barometrical disturbance is itself only local, and when the actual amount of difference between the extreme readings is less than half an inch, although the gradients for a short distance may be high.

The small trace of truth which the scale lettering in question contains is accounted for in the following way. On the average of a great many readings of the barometer, taken under various circumstances as regards the direction of the wind, it is found that in these islands the reading is highest when the wind is North-easterly, and, as has already been stated, this is when the air is coldest and driest. Consequently we have 'Very Dry' put down at the top of the scale, and according as the mercurial level rises from 'Change,' 29·5 inches, to its maximum height, we have 'Fair' or 'Set Fair' entered. Conversely, when the air is warm and moist and the wind South-westerly, the barometer is low, and so we have then the descending scale of 'Rain,' 'Much Rain,' and 'Stormy.' It is, however, quite a mistake to imagine that on any given day there is any certainty of the weather according with the description of it given by the word corresponding to the barometer height for that day.

There are yet other points of view from which the inutility, and in fact the absolute error, of these words may be indicated.

Let us take for example the word 'Change'; this is placed opposite the reading 29·5 inches, which reading is naturally supposed to be taken at sea level. If the barometer be removed to a station situated, say 500 feet, above that level, the corresponding reading will be about 29·0 inches, so that the whole scale will be half an inch out, and the error will be greater the more considerable the height of the station. The lettering is therefore again wrong, because it does not take account of the necessary reduction of the reading to sea level.

Once more, the range of the barometer is far greater in winter than in summer, so that the reading which corresponds to 'Fair' should be much nearer to 'Change' in summer than in winter. The lettering is still further in fault, therefore, as taking no account of this difference. The words are in fact little less than utter nonsense.

It is undeniable that there is more chance of strong wind when the barometer is low than when it is high, but this arises, not from the actual height of the barometer, but from the circumstance that cyclonic areas are usually much smaller than anticyclonic, so that when the barometer is low there is a greater probability of a steep gradient, from adjacent higher readings, existing in the neighbourhood and causing high winds, than when the barometer is high.

It, however, sometimes happens that the barometer in these islands will remain for a day or two below 29 inches: that is, below 'Stormy'; without any gale, because the area of low pressure is extensive and the gradients slight. An instance of these conditions has already been cited at p. 23, as having occurred March 9, 1876.

The question is often asked why gradients can be said *to be for certain winds?* The answer to this is very simple. Let us recur to fig. 2 (p. 34), and add to the curves and arrows already shown, a number of straight lines joining Holyhead to Valencia, Aberdeen, Skudesnaes, the Helder, and Brest. These lines (fig. 5) represent the direction of the gradients, and a glance at the chart will show that the winds, as regards their direction and force, bear a definite relation to these gradients in accordance with Buys Ballot's Law.

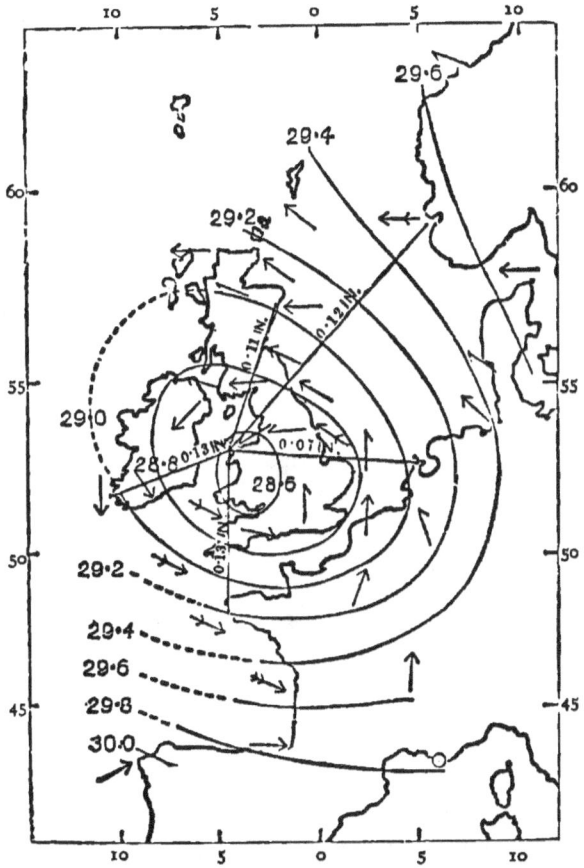

FIG. 5.—November 29, 1874, 8 a.m. Isobars and Wind, with Principal Gradients.

It will, however, be seen that as the form of the central isobar is oval, not circular, and as the lowest reading, that at Holyhead, is at some distance from the centre of the oval, the accordance of the winds with the gradients is not so exact as would be evinced in the case of a more perfectly circular storm. This is particularly the case as regards the relation of the winds at Scilly and Pembroke to the gradients between Brest and Holyhead. The centre of the disturbance really lies not far from Shrewsbury, and a gradient from Brest to Shrewsbury would be steeper than that from Brest to Holyhead, and the winds would be more nearly perpendicular to it than is the case in the chart. In fact I may exhibit this relation of the winds to the gradients by the following table:

GRADIENTS.

Stations	Amount	Direction of wind indicated	Actual winds
Valencia to Holyhead	0·13	NNW.	NW.$_8$ at Roche's Point
Aberdeen to ,,	0·11	E. by S.	E.$_7$ at Aberdeen SE.$_7$ at Leith
Skudesnaes to ,,	0·12	ESE.	SE.$_{10}$ at Scarboro' and Shields
The Helder to ,,	0·07	S.	S.$_3$ at Yarmouth S., at Cape Gris Nez
Brest to ,,	0·13	W.	WNW.$_{10}$ at Scilly W.$_9$ at Brest

It is needless to multiply examples. The chart shows a complete cyclonic circulation, and the gradients are taken on various sides of the centre. The entire condition furnishes a clear ocular proof of the statements that the gradients are nearly perpendicular to the isobars, while the wind is nearly perpendicular to the gradients, and therefore nearly parallel to the isobars.

Gradients.

From the way in which the direction of the gradient is described, we gather the direction of the wind indicated. When we speak of a gradient from Valencia to Holyhead being for North-westerly winds, we imply that the reading at Valencia is the higher of the two readings, *the station with the higher reading being always placed first*, and so a man standing midway between the two stations, with the lower barometer on his left-hand side, would face SSE. and have his back to the NNW.: the wind would therefore be North-North-westerly, which brings us back again to Buys Ballot's Law.

We have now learnt how the idea that the reading of the barometer at any time gives an indication of probable weather is controverted by the more correct conception of the value of gradients, but there is another idea which, though not absolutely misleading, is yet an unsafe mode of interpretation of barometrical readings. This is the idea that the motion of the mercury in the barometer is an infallible indication of coming weather. It is often asserted that if the barometer falls, say, at the rate of one-tenth of an inch per hour, a storm is certain to ensue at the place.

Depressions with slight gradients, moving quickly, will, however, cause the barometer to fall, at stations over which they pass, as rapidly as depressions with steep gradients which move slowly.

It is of course true that if an area of depression passes over us the barometer will fall till the centre has gone by, and then will rise, and that the rates of such fall and rise, for a given rate of motion, will bear a close relation to the amounts of the gradients along the path of the centre; but it is not by any means certain that because a barometer falls rapidly, there must be a gale at the place where this fall is observed. A very good instance of the fallacy of this opinion is to be found in the following circumstance, when the barometer fell with extraordinary rapidity over an extensive region, and yet hardly at a single station within that region was there any gale felt.

November 22, 1869, a very remarkable barometrical depression appeared over Western Europe. The actual fall of the mercury since 8 a.m. on the 21st had exceeded 0·9 inch over the entire district extending from Dover to Valencia and from L'Orient to Shields. This area is bounded by the meridians of 1° E. and 11° W. and by the parallels of 48° and 55° N., and its superficial extent is about 200,000 nautical square miles.

On this occasion, although there was such an enormous disturbance of equilibrium, there were absolutely no strong winds reported on our coasts during the day, excepting slight gales from North-west at Cape Clear, and from South-east at Yarmouth.

Outside our west coasts, however, on the edge of the disturbance, gradients were very steep, and there gales were felt between the 21st and 23rd, not only by the Atlantic steamers 'Scotia' and 'City of Brooklyn,' which were near the Irish coast, but also by the 'Inverness' and 'Foam,' which were further south. In fact, the last-named vessel, in the latitude of Cape Finisterre, was driven to the southward, out of her course, and suffered severely. At Corunna, however, and on the coast of Portugal, the gale was not felt seriously at all. In this case, therefore, a fall of the barometer of an inch at several stations was not followed by a gale at those stations.

March 9, 1876, the instance already quoted (p. 23), affords quite as striking an illustration. That morning at 8 a.m. the fall in the barometer within twenty-four hours had amounted to 1·10 inch at Wick, had exceeded an inch over the entire north and east of Scotland, and had been more than 0·8 inch over the rest of Great Britain, most of Ireland, and the north of France, yet the only wind of a force above 7, 'a moderate gale,' was at Rochefort, where a heavy Westerly gale was reported, and at the very centre of the disturbance, *where the barometrical fall was the greatest*, the force of the wind was the slightest, because the gradients were least.

This entire depression, like that mentioned just above, passed off without causing any gale worth notice in the United Kingdom.

We have learnt in this chapter that the distribution of pressure, as defined by the gradient, is the best guide, by the use of the barometer alone, towards a knowledge of the laws of wind motion, and, as we shall subsequently learn, of coming weather; but I must state that it appears that the force of the wind is not regulated *solely* by the gradients, though meteorologists have not yet determined what are the other agencies which influence it.

It is therefore evident that any attempt to foretell weather by the indications of the barometer at any one station, unsupported by observations of wind, clouds, weather, &c., must necessarily fail. The direction and force of the wind in no way depend on the actual barometer reading at the time.

We have, moreover, seen the meaning of the phrase 'gradients for such and such winds,' which are nothing more than the expression of the laws of wind motion in a practical form.

CHAPTER V.

CYCLONES AND ANTICYCLONES.

THE consideration of gradients, and of the effects of the distribution of pressure, brings us to the contrast of the two classes of atmospheric systems, anticyclonic and cyclonic, and the weather they bring with them.

Anticyclonic systems are marked by a very slow circulation of the air, or, in other words, by light winds, by low temperature in winter, great 'absolute' dryness of the air, at least at the centre, and consequent absence of rain, although fog is frequently very prevalent. These circumstances are accounted for, by some meteorologists, on the supposition that the air, in an anticyclone, flows out from the centre, and therefore must be supplied from the upper regions of the atmosphere by a descending current, which cannot possibly contain much moisture, owing to the very low temperature of the regions from whence it is drawn.

Cyclonic systems, on the contrary, are distinguished by the exact converse of the above conditions. The air circulates more rapidly, causing strong winds, and appears to flow in towards the centre, so that it must naturally be supplied from below and ascend in the centre. Cyclones bring with them, at least over a considerable proportion of the area they cover, a comparatively high temperature, much moisture, and consequently heavy rain. I say, over a considerable proportion of the circumference, because, in certain parts of a cyclonic system, the wind which is felt is often very dry. This is in the rear of the disturbance.

These statements refer to the winter; in summer the conditions are exactly reversed, at least as regards the temperature which accompanies the respective systems of disturbance.

Anticyclonic systems in summer produce our hottest weather, inasmuch as the air is so dry that no heavy clouds can be formed, so that the sun has full opportunity for exerting his heating power, while there is hardly any motion in the air to produce a cooling breeze.

Cyclonic systems in summer, on the contrary, bring cloudy weather, rain, and a reduction of temperature, the latter being mainly due to the density of the cloud covering which entirely intercepts the direct rays of the sun.

It is easy to illustrate these statements by examples. Let us take the instance of an anticyclone first, in fact the very day which we have already cited, February 4, 1874 (fig. 3, p. 37), which will serve as a very good illustration of a winter anticyclone. Fig. 6 exhibits the isobars as before, but the figures on the chart refer to temperature.

It will be perceived that over the central area, embraced by the isobar of 30·6 inches (see p. 36), the tem-

Cyclones and Anticyclones. 55

perature is lower than it is anywhere else in these islands, and that there is a fall in temperature noticed at every

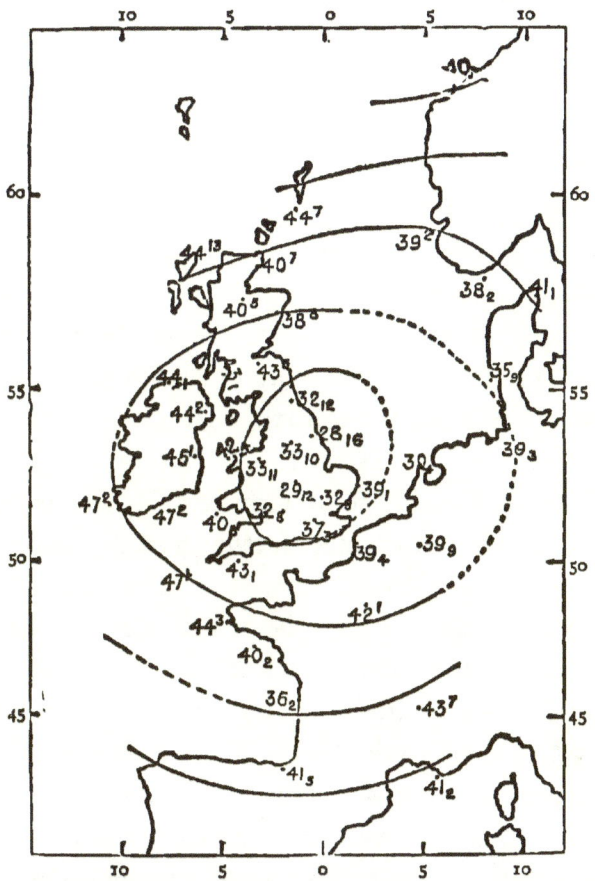

FIG. 6.—February 4, 1874; 8 a.m. Isobars and temperature, showing changes in previous 24 hours.[1] Anticyclone.

---

[1] The changes of temperature in figs. 6 and 7 are shown by small figures placed after the thermometrical readings. When these figures are *above* the readings, as at Wick, they indicate a *rise* of temperature; when they are below, as at Scarborough, they indicate a *fall*.

station, varying from 1° at Dover and Yarmouth to more than 10° in the north of England, and even 16° at Scarborough. In fact the cooling influence of the area of high pressure is noticeable over almost the whole map, except in Ireland and Scotland. In the north of the latter country, however, the rise in temperature is very marked, amounting to 8° at Aberdeen, and 7° at Wick and Sumburgh Head. This change of thermal conditions arose from the fact that, on the previous day, the north of Scotland had been the region of an independent anticyclonic system, as will be gathered from the examination of fig. 24, p. 86, but the barometer had fallen during the night, and the wind, though still light on our coasts, had shifted from N. to SW. both in Caithness and in the Lews. In fact this rise of temperature was connected with a cyclonic disturbance further northward, which is shown in fig. 26, p. 87.

As regards the force of the winds, a glance at fig. 3, p. 37, is sufficient to show that they are very light; in fact the only British or Irish station where they exceed the force of 5 of the Beaufort scale, a 'fresh breeze,' is Scilly.

The general dryness of the air is sufficiently evinced by the fact that rain was recorded for this morning from only five stations out of the fifty-one mentioned in the Reports, at three of them the amount being only 0·01 inch; the greatest quantity was at Scilly, and that was only 0·06 inch; nevertheless, the actual difference between the dry and wet bulb was but slight, showing that the atmosphere was near its point of saturation for the temperature. This circumstance is partly owing to the fact that the observations were taken at 8 a.m., so that the influence of the sun's heat had hardly made itself felt, and what moisture there was present was in the form of cloud or fog. Owing to this latter condition we find that the only region where the sky is clear was the north of Scotland, and that as we come south we meet with an entirely overcast sky at Leith, while fog and mist were reported all along the east coast of England, as being either prevalent at 8 a.m. or as having been so during the preceding twenty-four hours.

Frequently, though this was not the case in this particular instance, these fogs only last during the early part of the day and clear off about noon, so that the diurnal range of temperature during the existence of anticyclones is very great.

These conditions are more or less characteristic of the entire class of areas of high pressure, and, as we shall shortly proceed to show, the most striking feature of the weather which accompanies them is its permanency: *i.e.* the slowness with which changes succeed each other.

Let us now take the case of a cyclonic area (November 29, 1874), fig. 2, the very instance which has been already cited (p. 34), and examine into the features which characterised the weather which these conditions of pressure brought with them. This storm was so typical that the Weather Report and Charts for the day have been reproduced in Appendix A.

In the first instance we see (fig. 7), that over almost

the whole of the map the rise of temperature from the previous day (indicated by figures *above* the readings

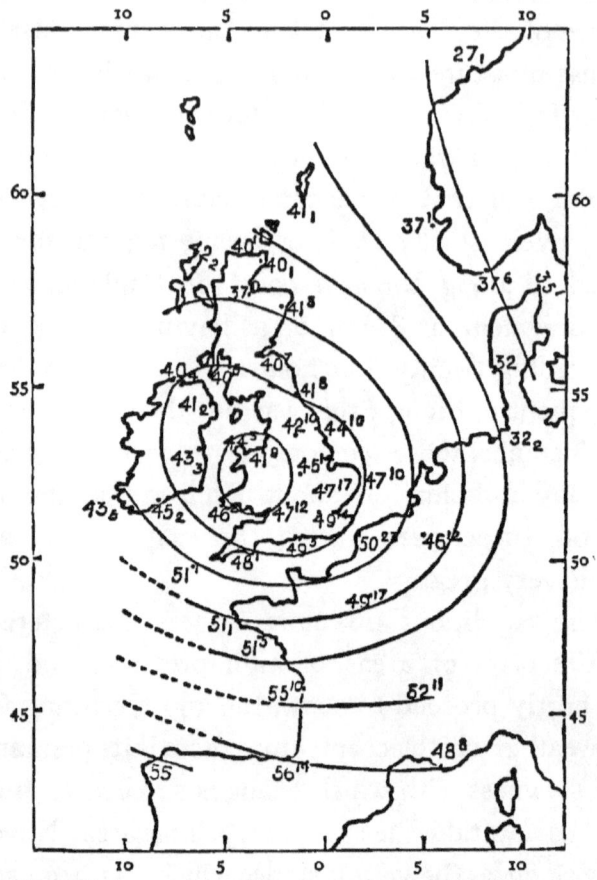

FIG. 7.—November 29, 1874; 8 a.m. Isobars and temperature, showing changes in previous 24 hours. Cyclone.

as before) has been very striking, especially at the south-eastern stations, where the increase has exceeded 10°, amounting even to 23° at Cape Gris Nez, near Boulogne. This region lies in front of the advancing disturbance, and is the district of the Southerly and South-east winds. The motion of this particular storm will be traced figs. 19-22, pp. 83-85.

In the rear of the storm, *i.e.* in Ireland and Cornwall, where the wind has shifted to North-west, we find that a fall (indicated by figures *below* the readings) is reported at all the Irish stations, as well as at those in the extreme north of Scotland. That this is due to the direction of the wind is sufficiently clear from the fact that at almost all the places where the cooling effect is observable, the direction of the wind is from some point to the Northward of West, being NE. at Donaghadee, N. at Valencia, NW. at Roche's Point, and WNW. in the south of England, where, although no fall of temperature from the reading at 8 a.m. on the previous day is reported, the rise is only 1° at Scilly and Plymouth, as compared with 10° at Dover and 14° in London.

As regards clouds and rain, the facts are still more striking. Out of the entire list of stations, only thirty-seven give the amount of cloud in their reports: of these twenty-three stated the sky to be 'entirely overcast,' thirteen others reported various proportions, varying from the above quantity down to that of six-tenths of the sky being covered, while at one solitary place, Hurst Castle, with a WNW. wind, the sky was half clear.

The rainfall is equally marked. During the twenty-four hours preceding the epoch for which the chart is drawn, rain fell at every station in these islands and France, whence we have reports, except Nairn, and, as to its amount, at four stations this exceeded an inch, and at ten others it was more than 0·7 inch.

It may therefore be said that areas of low pressure are accompanied by a high temperature, especially in the front of the storm, a large amount of moisture in the air, and consequent rainfall and prevalence of cloud. The instance which has been given, however, although it illustrates very well many of the features of a cyclonic disturbance, does not exhibit all the circumstances which may be observed in a given storm as to the march of the different phenomena within the system itself. One of the most striking characteristics of a cyclonic storm is a sudden shift of wind which takes place between SW. and NW., accompanied frequently by a heavy squall and a shower, together with an almost instantaneous fall of temperature, and this is entirely absent in the case in question, as will be explained on p. 85. To take a very remarkable instance, we find that on February 12, 1869, a cyclonic storm passed up the English Channel; the lowest barometrical reading being recorded at Falmouth at noon, when the wind flew from SW. to WNW., and the temperature fell 6°·2 in a few minutes. At Kew similar changes took place at 4.30 p.m., but in a very aggravated form: the shift of wind having been from WSW. to ENE., and the fall of the thermometer 11°·6 in about five minutes.

The storm of Sunday, March 12, 1876, will be fresh in the memory of all residents in the south of England. In this case, which is too recent to discuss at length in these pages, the shift of wind at Kew was from SW. to N., and the fall of temperature 8°·5 in 30 minutes, accompanied by a heavy snow shower.

I have already said that one great distinction between cyclones and anticyclones, is that the former move, and the latter are usually nearly stationary, and as the usual motion of a cyclonic system is from West to East across these islands, it will be interesting to see what we can learn from the continuous records at our self-recording observatories of the actual changes which will occur at a station during the passage of a storm in that direction.

This will naturally differ according as the station lies to the North or the South of the centre of the storm, and, as it very rarely happens that the storms are truly circular, it is obvious that the shifts of wind and changes of weather will not all be perfectly regular. In fact the extraordinary changes of February 12, 1869, just cited, afford a very good proof of the unequal distribution of the gradients in such a system of disturbance.

Let us first take the most common case in these islands, that of a station lying on the southern side of the path of the centre of depression, moving from West to East across central England, and that we are situated in London. We shall experience first the phenomena belonging to the front of the system: the appearance of cirrus clouds, 'mare's tails,' in the sky, then the South-easterly winds, the great rise of the thermometer and excessive dampness, the sky becoming gradually overcast and the setting in of mist and rain, the barometer falling persistently, while scud begins to drift from the southward. As the system advances the barometer continues to fall, the wind veering through S. to SW., and rain falling steadily. As soon as the wind passes the SW. point and draws to W. or NW., the barometer begins to rise, often with a sudden jump, and the temperature falls, with a very heavy shower of rain, possibly turning to hail, after which the air becomes much drier, and the sky clears with a NW. wind, which as a rule soon dies down. After such a disturbance as this we have frequently a smart frost at night, inasmuch as the great dryness of the air has allowed great radiation of heat from the ground.

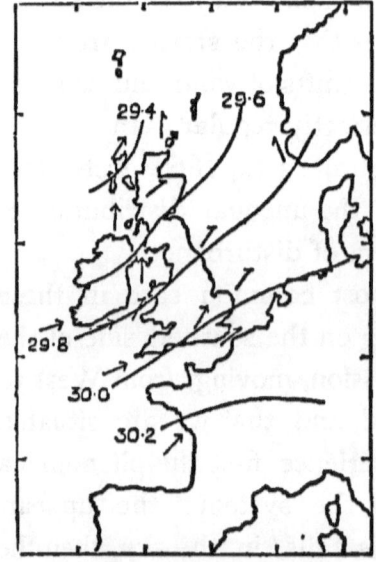

FIG. 8.—March 27, 1874; 2 p.m. Storm passing north of Valencia.

It is not easy to find an instance which exhibits the entire course of changes just described, but the table, Appendix B., will illustrate several of the points.

The table represents the hourly readings of the barometer, thermometer (dry and wet), anemometer (direction and velocity), rain-gauge, and the vapour tension from the continuous records at Valencia on March 26 and 27, 1874, when the centre of a cyclonic storm passed to the north-

Cyclones and Anticyclones. 63

ward of that observatory. At 2 p.m., March 27, (fig. 8,) the centre of the storm lay off the North-west coast of

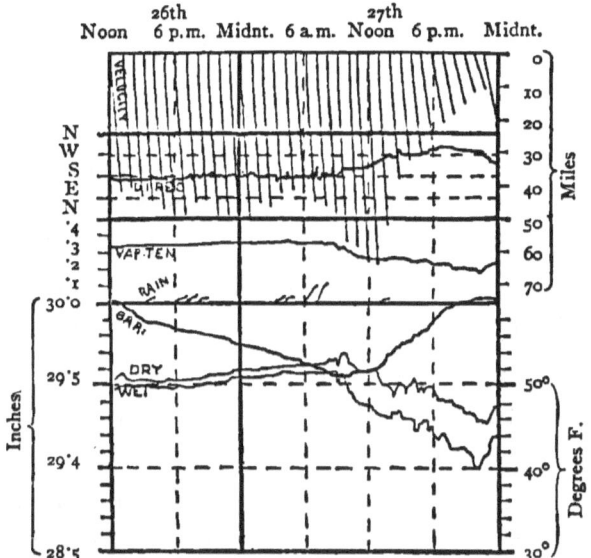

Fig. 9.—March 26 and 27, 1874.   Automatic records at Valencia.

Scotland, outside the Hebrides. The diagram (fig. 9) exhibits the course of the curves.[1]

---

[1] These diagrams require explanation. They are copied from the plates in the Quarterly Weather Report, and exhibit the continuous records of:

1. The barogram or barometer curve, marked *bar.* The scale for this is given in the lower left-hand corner. It is in inches, and the readings, though corrected for temperature, are not reduced to sea level, and range over 1½ inches.

2. The thermograms, marked respectively *dry* and *wet.* The scale for these is in the lower right-hand corner, and ranges over 30°.

3. The curve of tension of vapour, marked *vap. ten.*, calculated from the two thermograms, dry and wet.

4. The rain in hourly amounts of fall. These are measured upwards from the base line of the scale which serves for vapour tension and rain,

The barometer fell steadily for twenty-three hours, until 9 a.m. on the 27th, and then rose in twelve hours nearly to the same height as it had held before. Temperature and vapour tension remained singularly steady during the afternoon of the 26th, rising very slowly during the night, and reaching a maximum at 8 a.m. nearly simultaneously with the epoch of the lowest barometer readings. The dry thermometer then began to fall briskly, and by 9 p.m. had sunk $5°·7$. It will be observed that, owing to the influence of the storm, the temperature was higher during the early morning hours of the 27th than it was during the afternoon of either the 26th or 27th, and yet 2 p.m. is usually about the hottest part of the day. While it lasted, therefore, the storm entirely disturbed the regular diurnal march of temperature. The secret of this disturbance of temperature lies in the wind.

During the whole time that the barometer was falling the wind remained nearly constant between SSE. and S., blowing a stiff gale, and a little rain fell. When the centre of the storm was approaching, the rain set in again at about 4 a.m., and became heaviest just at

and is on the left-hand side, above the barometer scale ; it is in tenths of an inch, and extends over half an inch.

5. The wind direction, marked *direc.* This is on a special scale shown at the left-hand side, which represents a paper wrapped round a cylinder, and then laid flat. N is at the top, then come successively E, S, W, and N again.

6. The wind velocity, marked *velocity*, in hourly amounts, measured by the scale of miles up to seventy, shown on the right-hand side.

The time scale is shown at top and bottom.

8 a.m., when the barometer was nearly at its lowest level and the wind was about veering towards SW. At the hour of the lowest barometer reading we find the first trace of Westing in the direction of the wind, and at the same hour the velocity shows a sudden increase, the rain still continuing. The sky then cleared, the barometer rose briskly, and the wind increased till noon, when it blew a heavy gale from SW.byS. It subsequently veered pretty quickly to WNW. with a slight shower of rain at 3 p.m., and fell very light after sunset, at 7 p.m.

The courses of the wet bulb thermometer curve, and of the curve of vapour tension, illustrate very clearly the contrast between the Southerly and North-westerly winds as regards the humidity of the air. As long as the wind remained to the Eastward of South the curves of the two thermometers remained close to each other and the curve of vapour tension rose steadily, but very slowly. At the hour of the minimum barometer reading, we notice the curves going apart, and for the remainder of the period they pursued courses diverging more and more from each other, indicating, what is also shown by the descent of the vapour tension trace, that the air was becoming drier and drier according as the wind became more Northerly.

Owing to this extreme dryness of the air of the Northwest wind, in the rear of a storm, it is believed by some meteorologists that this wind is a downrush of air from the upper regions of the atmosphere, which is necessarily very dry, owing to the low temperature to which it has been exposed.

The case just cited, though it exhibits some of the phenomena with great distinctness, is not one which can be made clear to the reader as an instance of a storm passing to the northward of a station, inasmuch as the system to which the gale in question belonged was so extensive that the limited area of our charts would not exhibit the conditions in a convincing manner. Fig. 10,

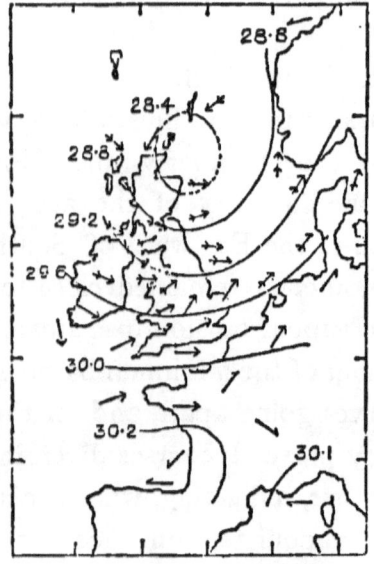

FIG. 10.—October 21, 1874; 8 a.m. Storm passing north of Aberdeen.

however, is an indisputable case of a storm passing on the northern side of Aberdeen, and close to it, October 20-21, 1874, and the course of the curves at that observatory is shown on fig. 11. The actual hourly readings will be found in Appendix C.

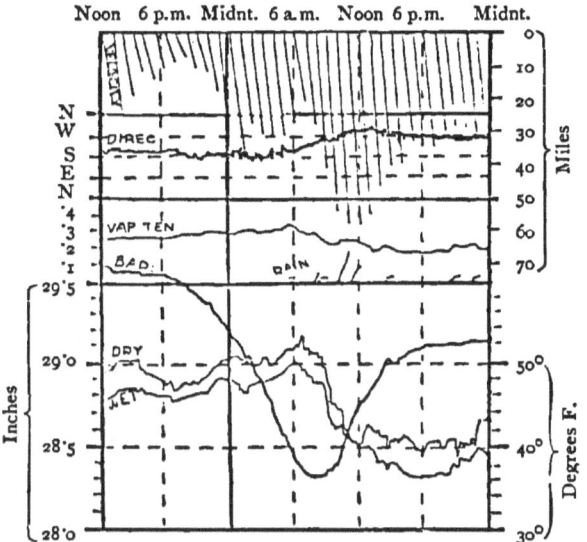

FIG. 11.—October 20 and 21, 1874. Automatic records at Aberdeen.

The fall of the barometer was slow at first, but then increased in rapidity, and in twelve hours the mercury sank to the extent of 1·14 inch. It then rose in the next twelve hours as much as 0·8 inch, and for a portion of the time the rate of rise was more rapid than that of the previous fall had been. It will be noticed that the epoch of the maximum rise of the barometer was also that of the shift of wind through three points to the Northward, as well as of the strongest wind and the heaviest rain. This is the heavy shower of rain, coincident with the shift of wind to the North-west, of which mention has previously been made, p. 60. Temperature and humidity do not follow a similar course respectively as they did in fig. 9, as both curves are far less steady in their course than on the former occasion. We see, however, that the diurnal march of temperature is again entirely disarranged, the thermometer remaining pretty nearly at the same temperature throughout the night of the 20th, and the sudden fall of both temperature and vapour tension taking place nearly simultaneously with the shift of wind above noticed, and at the time when the natural rise of day temperature ought to be most brisk, viz., about 10 or 11 a.m.

The wind commenced at about SSW., and blew pretty steadily from that quarter for twenty hours, its velocity being at first slight, and increasing to that of a stiff breeze at midnight. When the barometer was at its lowest and beginning to rise, the direction of the wind rapidly shifted to WNW., and its velocity more than doubled itself in three hours, rising from 24 to 55 miles an hour. The velocity of a strong gale, however, only lasted for four hours, from 9 a.m. to 1 p.m. The absence of rain is very noticeable during the early period of the gale; the reason for this absence can be seen from the fact of the great distance between the wet and dry thermograms; in fact no rain of any consequence fell excepting just at the period of the lowest barometer, already noticed.

It will be seen from figs. 48 and 49, p. 140, 141, that this storm, which was a very severe one in other parts, passed just to the northward of Aberdeen.

Taking now the second case, that of a cyclónic system passing on the southern side of the station, we have a very good instance of the course of the phenomena in the records of Falmouth for February 1 and 2, 1873. The hourly values, as before, are given in Appendix D, and the instrumental records are shown on fig. 13, while the conditions of pressure at 8 a.m. on the 2nd are shown on fig. 12.

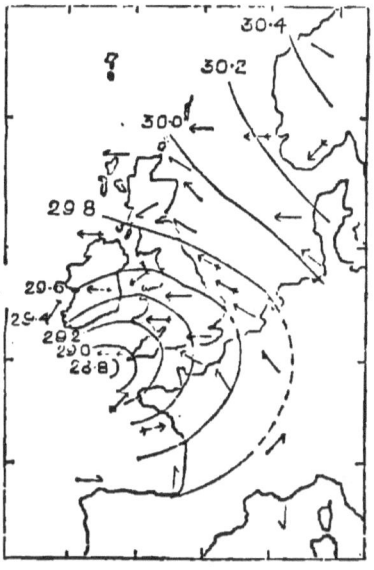

FIG. 12.—February 2, 1873; 8 a.m. Storm passing south of Falmouth.

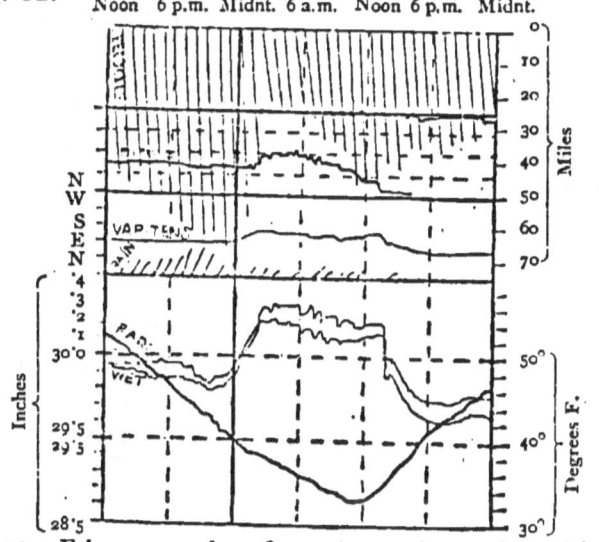

FIG. 13.—February 1 and 2, 1873. Automatic records at Falmouth.

Here, too, as in the previous case, we have a great dip in the barometrical curve, which falls very regularly for twenty-four hours until noon on the 2nd. It then rises more rapidly than it had fallen before, and at midnight on the 2nd attains to within a quarter of an inch of its previous height.

Temperature shows very remarkable changes during the night of the 1st and 2nd, rising 10° between 10 p.m. and 4 a.m. It then continues high until 2 p.m. on the 2nd, when it falls 6° in a few minutes, a change nearly as sudden as that at the same observatory on February 12, 1869, already noticed (p. 60).

As before, we find the explanation of these changes in the direction of the wind. Up to midnight on the 1st this was steady at ESE. It then veered slightly to SE., and even a couple of points further. The highest temperature, already noticed, coincided with the direction S.byE. As the centre drew nearer, the vane backed smartly to E.byS., E.byN., and NE., and the sudden fall of temperature marked a shift from NE.by N., to N.byE. The shift went on further, and the period closed with the wind about NNW.

The extreme force of the wind did not coincide precisely with the period of the lowest barometer reading, as it was far greater a little before midnight on the 1st than at any other time. This affords a further proof, if such is needed, that the violence of a gale does not depend on the actual height of the barometer.

Vapour tension showed a steady curve, following that

of temperature, until the appearance of the Northerly and North-westerly winds, when it sank in a marked way, thus again corroborating the former statement as to the dryness of winds from that quarter.

The rainfall was very remarkable; it continued almost without intermission from noon on the 1st till 5 p.m. on the 2nd, being heaviest shortly before the period of the strongest wind. The total amount which was collected during the twenty-four hours, ending with noon February 2, was no less than 1·275 inch, and it will be noticed that the rain gradually lessened as the wind backed through N., and ceased entirely when the direction was NNW.

This persistence of rain with the Easterly and North-east winds in the front of a cyclonic depression is a very striking characteristic of these disturbances. It is an unmistakeable proof that North-east winds are not always dry, and the only reason that this fact is not more frequently observed, is, as will shortly be shown, that the occurrence of these winds in the front of cyclonic disturbances is a comparatively rare phenomenon in the British Isles, for the storms usually pass by us on the northern side, and at first give us South-easterly, not North-easterly winds.

These instances are amply sufficient to show the way in which we are to interpret a couplet well known to sailors, which expresses Dove's Law of Gyration referred to at p. 25 :

> When the wind shifts against the sun,
> Trust it not, for back it will run.

Under certain conditions a backing of the wind is quite *en règle*, and is simply an indication that a cyclonic disturbance is passing on the southern side of the observer. The actual motion of the disturbance of February 1 and 2, 1873, was along the north coast of Brittany, and therefore to the south of Falmouth.

The rule as regards the veering or backing of the wind in connection with atmospherical disturbances may be thus stated.

If the observer supposes himself at the centre of a cyclonic system in the northern hemisphere and moving with it, the wind at all stations which he passes will *veer* if the station be on his *right-hand side* and will *back* if the station be on his *left-hand side*.

This is made clear by the following diagrams (fig. 14, *a, b, c, d,*) in which I. II. III. indicate the successive positions of a cyclonic system, and the large arrows represent different directions of its motion. A is always on the right-hand side of the centre, B on its left-hand side. The successive directions of wind experienced at each station are shown by the figures 1, 2, 3, and 1' 2' 3'. In every case it will be seen that the shift from 1 to 3 is *veering*, and from 1' to 3' is *backing*, whatever the first direction of the wind may have been.

The reason that a backing wind is considered dangerous is, that when the wind backs from NW. towards SW., S. and SE. at a station A in fig. 14, it indicates that the SE. wind of the North-eastern side of another area of low pressure is approaching the station which has

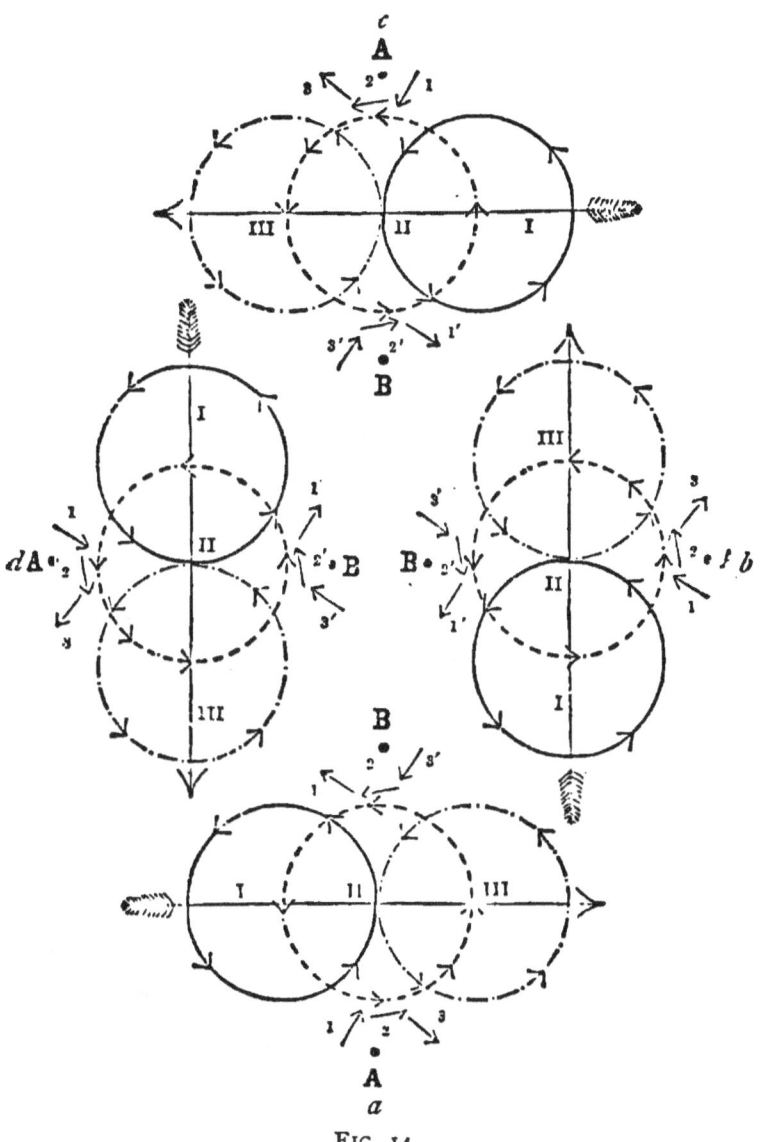

FIG. 14.

been experiencing the NW. wind of the South-western side of the previous area of low pressure, and which is passing away to the eastward, so that the wind merely backs until the station is fairly under the influence of the new disturbance, when it begins to veer again, if the centre of the depression passes to the northward of the station, as is usually the case in these islands.

There is one point in connection with cyclonic disturbances which is gradually attracting more and more attention, and that is the appearance of secondary eddies in connection with the larger areas of barometrical depression. These smaller systems are usually less perfectly developed than the larger ones, and they manifest themselves in general on the southern sides of the latter. Accordingly, while they cannot exhibit any Easterly winds of much force, (owing to the fact that pressure is lower on their northern than on their southern edge), they intensify the Westerly winds on the extreme Southern edge of the original disturbance by increasing the gradients in that part of the system.

It is probably owing to these circumstances that Easterly winds are so rare in our storms. Investigations into the distribution of pressure over the earth's surface have shown that there is an almost constant deficiency of pressure in the neighbourhood of Iceland, and so probably most of the depressions which cause our storms are simply secondary to a vastly more extensive area of depression over the North Atlantic.

It will be interesting to give a few instances of these

'satellite' depressions (as the French call them), to illustrate what I have been saying, and fig. 15, for January 3, 1874, shows us on a small scale the conditions to which

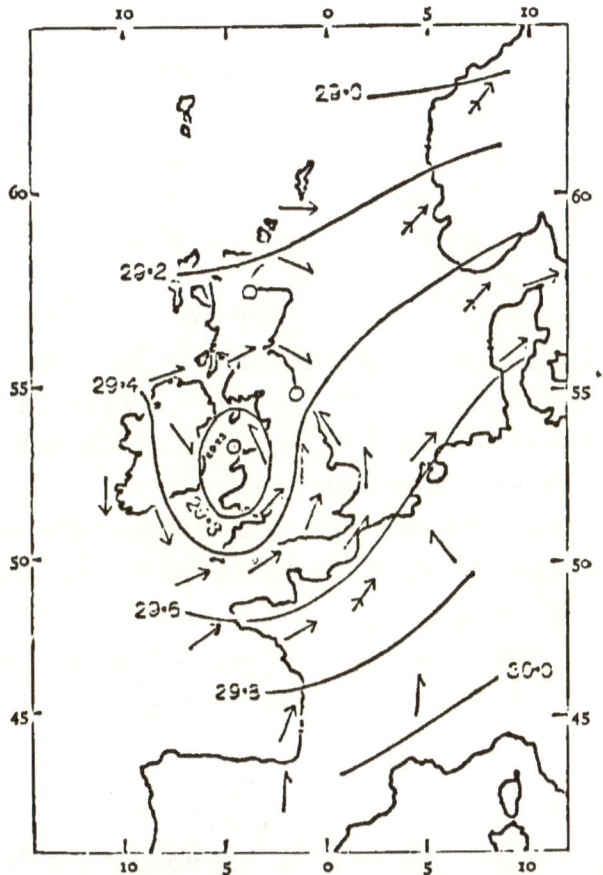

Fig. 15.—January 3, 1874; 8 a.m. Subsidiary depression.

allusion has just been made as characterising many of our gales. We see that there is a small cyclonic disturbance with its centre near Holyhead, and that the circulation round it is very imperfectly developed, inasmuch as there are no Easterly winds at all. The reason of this deficiency is at once visible when we look at the northern part of the chart. The reading of 29·23 inches, at Holyhead, which is at the centre of the subsidiary depression, appears again over Caithness, in the isobar of 29·2, and to the northward of it again we see that of 29·0 inches, so that evidently readings must be still lower if we went further north. As the reading both at Holyhead and at Wick is about 29·2, and as the highest reading between these points is 29·3 inches, it is impossible that there can be a gradient of any extent for Easterly winds over the intervening region, and so those winds do not appear. On the other hand the circulation round the west, south, and east sides of the subsidiary depression is clearly shown.

It is not often, however, that we find the secondary depressions so clearly marked as in fig. 15. The chart for October 9, 1874, gives us their more common character (fig. 16). They manifest themselves simply by a loop in some of the isobars, and their effect is, as already described, to reduce the gradients on the side turned towards their primary, and to increase them on the opposite side. We could hardly have a better example of this than in fig. 16. The strong NW. wind at Stornoway, and the SW. gales in the Skagerrack and in Denmark, show that the conditions for strong winds prevail generally, but over southern Scotland and central England nothing but light Westerly airs are reported. On the other hand, in

Cyclones and Anticyclones. 77

the Channel and in France the isobars are crowded together, and heavy gales are the result.

Another instance is represented on fig. 17, for October 22, 1874, but here it is the North-westerly winds of the primary that are reduced in force, on account of the fact of the secondary system manifesting itself on

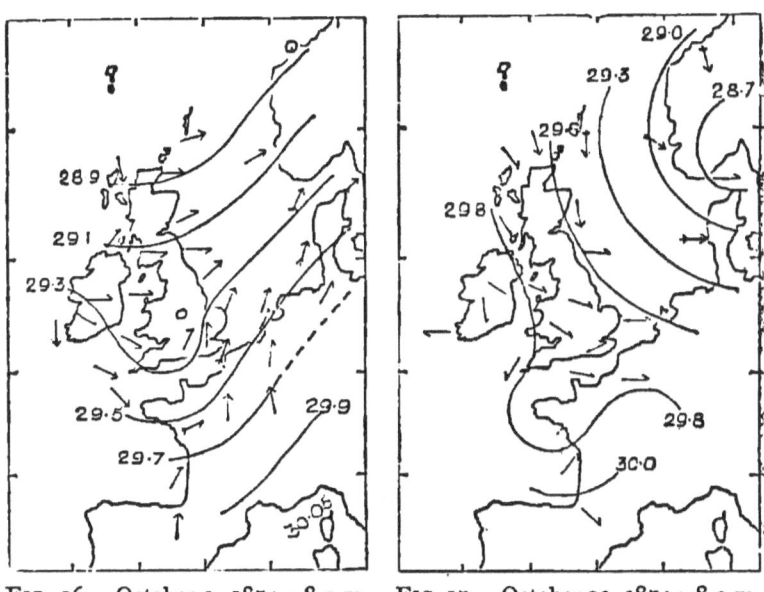

FIG. 16.—October 9, 1874; 8 a.m. FIG. 17.—October 22, 1874; 8 a.m.
Subsidiary depression. Subsidiary depression.

the south-west side of its principal. Owing to the apparent equality of pressure over the Bay of Biscay there are no gradients of consequence there, and therefore no gales in connection with the smaller system, but so far as direction is concerned, the circulation round the smaller depression is very clearly marked.

At times it is hard to say which of the depressions is

the primary, and which the satellite, as there is apparently not much difference in size between them. This was the case on January 27, 1872 (fig. 18), when we see

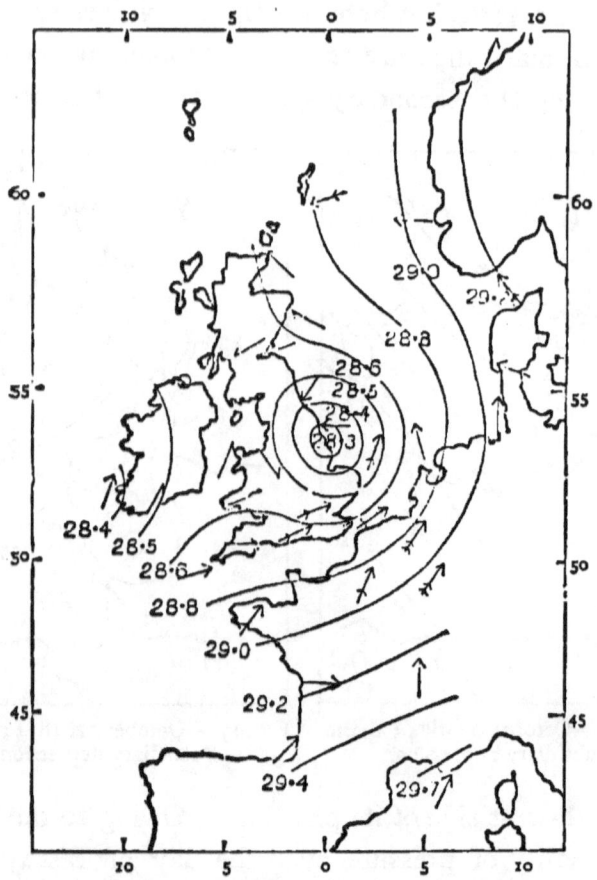

FIG. 18.—January 27, 1872; 8 a.m.   Coexisting depressions.

one depression, with central readings as low as 28·3 inches, near Scarborough, while another shows itself off Valencia. The result of the interference of these two systems is to produce a comparative calm between them, the winds being very light over St. George's Channel, while strong gales are blowing wherever this interference does not exist.

In treating of the interpretation of weather charts and of storm warnings, I shall show the use which can sometimes be made of these subsidiary cyclonic systems.

In this chapter we have considered the contrast between cyclonic and anticyclonic systems, or areas, as they are indiscriminately termed, and the character of the weather which accompanies each respectively.

We have learnt that while anticyclonic areas move slowly, cyclonic areas travel more or less rapidly over the country, and moreover that their passage past a station presents marked features of difference, according as the station lies on the right or left side of the path of the centre. Lastly, we have found that cyclonic systems are not always isolated, but occasionally have what may be called satellite systems in connection with them which exert a material influence on the gradients, and consequently on the winds belonging to the original disturbance.

CHAPTER VI.

THE MOTION OF STORMS AND THE AGENCIES WHICH APPEAR TO AFFECT IT.

BEFORE we treat of the motion of storms it must be admitted, as will appear in the next chapter, that meteorologists are not at all agreed as to what the real causes of that motion are. It has not been satisfactorily determined to what intent the storm moves forward as a whole, or to what extent the advance observed is only due to the continued reproduction of the same phenomena at successive points, as is the case in wave motion.

The fact that storms are swept onward over the earth's surface with a motion of translation in addition to their own rotatory motion has, however, long been known. The earliest notice of it which we can discover is an entry on a map of Virginia, published in 1747 by Lewis Evans, to the effect that 'all our great storms begin to leeward.' Franklin, in 1760, followed in the same strain, but it appears that his attention had been caught, at an earlier period, in 1743, by the fact of his being prevented, by the clouds brought by a hurricane,

The Motion of Storms.

from observing a lunar eclipse at Philadelphia, while the eclipse was seen at Boston, which lies further to the North-eastward, before the storm came on. The reader hardly needs to be reminded that hurricanes and cyclonic disturbances in general are accompanied by an abundance of cloud. Along the eastern coast of the United States, therefore, the advance of storms from south-west towards north-east has long been a recognised fact.

The direction of the motion is different in different parts of the earth. Of storm tracks which are well made out, those of the West India Hurricanes, in the earlier part of their course, are, speaking in general terms, from east-south-east to west-north-west: in the Indian Ocean the Mauritius Hurricanes advance at first from east-north-east to west-south-west, but both of these classes of storms sometimes recurve subsequently at a sharp angle, and advance towards the eastward. The Typhoons of the China seas move towards the coast from the eastward.

Over these islands, however, the motion of storms is not nearly so constant in direction as in the instances which have been cited, and although as a general rule it may be said that our storms travel from the westward, there are many exceptions to this rule, and in fact storms can, and do, move from every point of the compass, but a motion from an easterly point is extremely rare in this part of the world.

It will be best to consider the less usual directions of motion in connection with the causes which appear to give rise to such variations in the usual course of affairs, and I shall also offer some remarks on the motion of anticyclones, and show what a contrast this presents to the behaviour of cyclonic systems.

As regards the rate of motion I may premise that this varies very greatly, some storms travelling very slowly or even remaining apparently steady, while others are propagated with great rapidity. The rate of motion of West India hurricanes ranges from ten to fifteen or twenty miles an hour in their early stages, but becomes accelerated in the latter portion of their path, while in these islands the motion has attained the prodigious speed of fifty miles an hour, February 12, 1868, an instance already cited, p. 60, and even seventy miles an hour, December 16, 1869, a velocity nearly equalled on the night of November 10–11, 1875, as well as March 12, 1876.

It must be clearly understood that the rate of translation of the storm *has no relation to the velocity of the wind in that storm*, for the West India hurricanes exhibit almost the strongest winds of which we have any knowledge, and they travel slowly, while the rapidly moving storms just cited were not exceptional as regards their violence.

However, one of the greatest difficulties which meets us in the issue of Storm Warnings is our almost total ignorance of the rate at which any given storm is travelling, until it has already moved over a considerable tract of country.

The Motion of Storms.

It is easy to cite instances of the motion of storms, and I shall select two of the cases which have already been employed as illustrations of cyclonic disturbances, and trace their progress across the area of our weather maps, paying attention mainly to the isobars and the wind.

Let us first take the storm of November 29, 1874. The earliest unmistakeable signs of its approach (fig. 19),

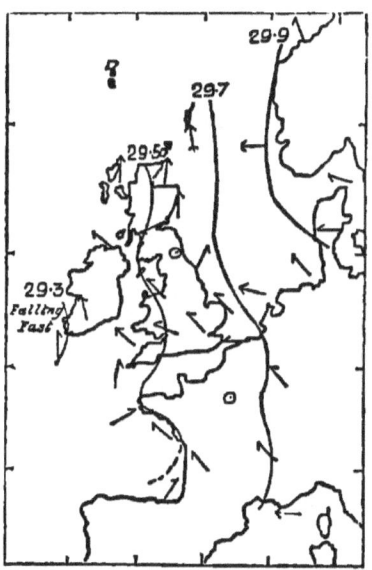

FIG. 19.—November 28, 1874;
8 a.m. Approaching depression.

were at 8 a.m., November 28, when a rapid fall of the barometer at Valencia, with the Southerly wind, and the course of the isobar of 29·3 inches, show that there must be an area of lower readings at sea, outside the coast.

Over the greater part of England the direction of

the wind is South-easterly; a very general phenomenon on the approach of a serious storm, owing apparently to the in-draught of air towards the region of diminished pressure.

The next chart, for 6 p.m. on the same day (fig. 20), shows the central isobar (of 29·0 inches) over the south

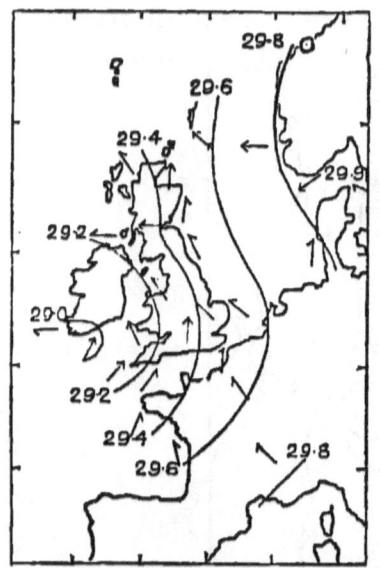
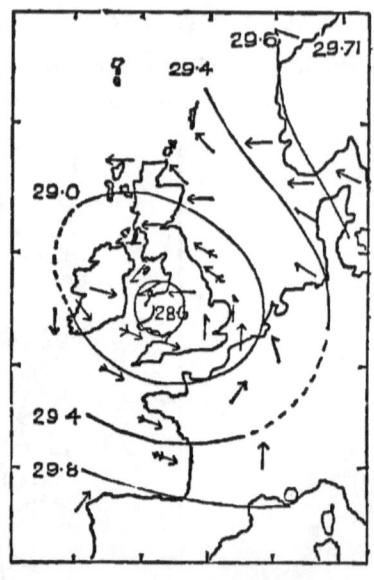

FIG. 20.—November 28, 1874; FIG. 21.—November 29, 1874;
6 p.m. Depression advancing.   8 a.m. Centre over Wales.

of Ireland, while the SE. winds over England have veered to SW., and the isobars, previously running nearly north and south, show a marked curvature. Even in Norway pressure has given way, the isobar of 29·8 inches having taken the place of that of 29·9 inches.

During the night the storm made rapid progress, and at 8 a.m. (fig. 21), we have the conditions already

The Motion of Storms. 85

described at p. 34, the centre of the storm lying near Holyhead, and the influence of the depression extending over the whole of western Europe.

At 6 p.m. on the 29th (fig. 22), we find the centre of the storm near Newcastle, with a reading of 28·4, while the general course of the isobars follows that of an ellipse, with its longer axis stretching east and west. The Southerly winds have nearly entirely disappeared, and the gales are from East in Scotland, and from West over England, without any amount of North-west winds worth notice.

If we were to pursue this disturbance further, we should discover that the oval shape of the isobars was caused by the approach of a second cyclonic area towards Ireland, which checked the rise of the mercury and reduced the gradients in the rear of the former system, and thus hindered the development of the North-westerly winds, a feature to which allusion has already been made at p. 60. In fact the chart for the next morning exhibited two separate areas of depression, one over Ireland, the other over the North Sea.

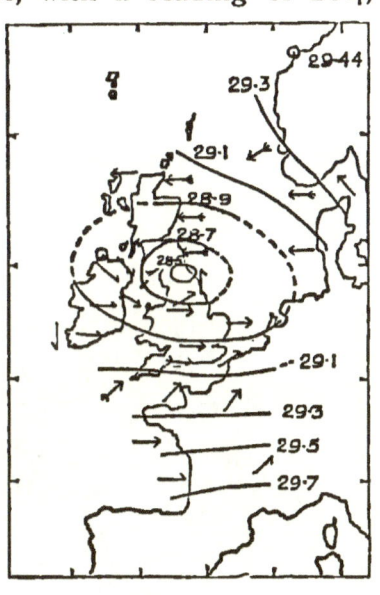

FIG. 22.—November 29, 1874; 6 p.m. Centre near Scarborough. Depression passing off.

while the latter area showed evident signs of being filled up.

It is not necessary to quote other instances to prove the fact of the motion of cyclones, and we shall now proceed, by way of contrast, to consider the motion, or rather the comparative absence of motion, in anticyclones, and shall select the period at the beginning of February 1874. It is not necessary to give more than the daily charts for 8 a.m., as the changes from day to day are very slight.

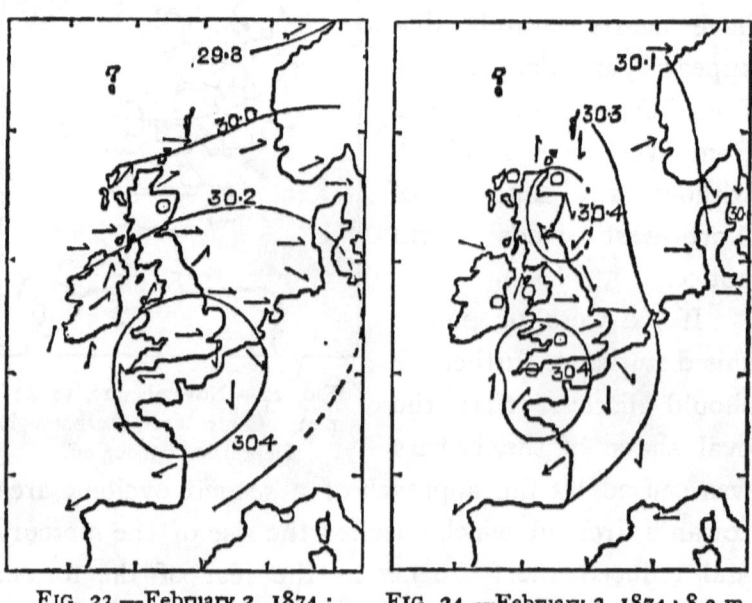

FIG. 23.—February 2, 1874; 8 a.m. Anticyclone.

FIG. 24.—February 3, 1874; 8 a.m. Anticyclone stationary, but changing in shape.

Fig. 23, for February 2, shows an area of readings above 30·4 inches over the Channel. In fig. 24 this

The Motion of Storms. 87

has hardly changed its place, but a sudden rise of the barometer has taken place over the north of Scotland, so that we have two independent centres of high pressure over our limited area. In fig. 25, for the 4th, the secondary anticyclone has vanished, and we have the conditions already cited at p. 36 as a typical anticyclone,

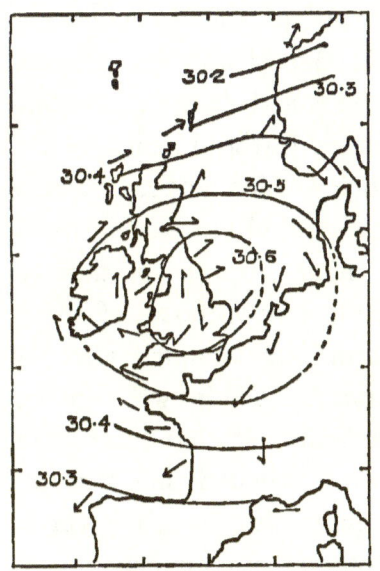
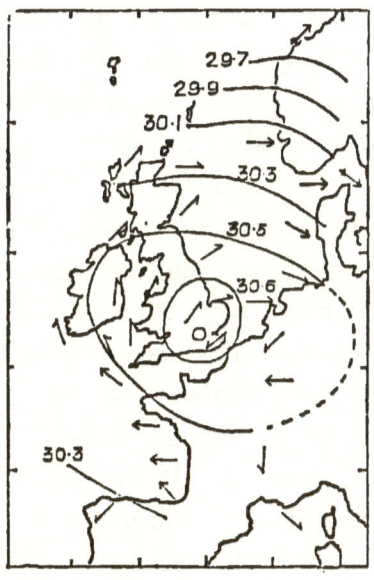

FIG. 25.—February 4, 18´4; 8 a.m. Original shape regained.

FIG. 26.—February 5, 1874; 8 a.m. Anticyclone disappearing.

and at the same time the effects of a cyclone outside the Shetlands are just felt on the coast of Norway, where the barometer has begun to give way, accompanied by a stiff South-west gale at Christiansund.

The last chart of the series, fig. 26, shows us the central isobar of 30·6 inches over the eastern counties, while all over the North Sea strong Westerly winds have appeared, owing to the advance over Norway of the cyclonic disturbance just mentioned.

In this case, therefore, which is a fair sample of the behaviour of areas of high pressure, it is evident that the persistence of the conditions of high barometrical pressure over the south of England was as marked as the instability of the conditions of low barometrical readings was proved to be in the preceding instance. Anticyclones, however, do change their position, but it is needless to pursue the subject further in this place.

Before treating of the agencies which appear to rule the motion of cyclones, it may be well to say a word or two on the influences which appear to modify that motion, at least in the neighbourhood of these islands. We find that the storm path is much affected by the contour of the country. It is well known that the west coasts of Ireland and Scotland are bold and rugged. It has sometimes happened that a storm has advanced from the Atlantic to the coast of Kerry, but has returned, and passing out to sea again, has moved northwards along the coast until it found the opening of Donegal Bay, and has then crossed Ireland along the tract of low land stretching from Ballyshannon to Dundalk, to the Irish Sea, where the centre often appears to be arrested for a while: or else the disturbance, clearing Ireland altogether, and crossing Scotland south of the Grampians, has passed out to the North Sea. In fact we have already seen, from the phenomena of the Hallsberg

The Motion of Storms.

Tornado, p. 25, that the stratum of air affected by a storm is frequently very thin as compared with the entire depth of the atmosphere. We can therefore easily understand that eddies in the stratum of air which is in immediate contact with the surface of the ground may be very much affected in their character and motion by the irregularities of that surface, which have a tendency to turn them aside.

It need hardly be pointed out (as suggested to me by my friend Prof. John Purser of Belfast) that the susceptibility of storms to such an influence, exerted by the contour of the country, bears some resemblance to the behaviour of smoke-rings, which are deflected from their course by obstacles placed in their path, although they may not actually come in contact with such obstacles. Remembering that the depth of storms is insignificant compared with their lateral extent, we may to a certain extent compare cyclonic disturbances to smoke-rings, although the constitution of the two phenomena is widely different.

As regards the actual direction of motion of cyclones, when we come to examine a series of charts extending over a wide area, like those of Captain Hoffmeyer, for Europe and the Atlantic, it is found that the cyclones frequently have a tendency to move round the anticyclones, but it is almost impossible to follow out these motions fully when we can deal only with the limited area of our own charts. It is, however, from this mutual action of the areas of high and low pressure on each other that we gain some notion of the coasts which are likely to be visited by a storm, of the direction which that storm will take, and of the quarter whence the wind in that storm will blow hardest.

If we find the readings highest over Ireland, the depressions will sweep down over Scandinavia or even over the North Sea, in a direction from the northward to the southward, giving us Northerly gales along the east coast of Scotland and England, owing to the fact that the steepest gradients will be on their western sides. These areas of low pressure will follow each other until by some means or other the excess of pressure in the west has been reduced.

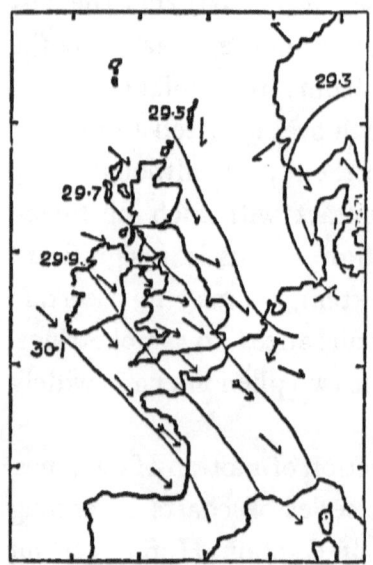

FIG. 27.—November 17, 1874; 8 a.m. Pressure highest in south-west. Depression disappearing over Denmark.

Figs. 27–30 afford very good examples of the above conditions. We start on November 17, 1874 (fig. 27), with readings highest in the south-west, general North-west winds, and a definite depression existing over Denmark, marked by the isobar of 29·2 inches. At the same time a backing wind at Stornoway and the SSE. wind at Christiansund show that a fresh disturbance is not far off.

The Motion of Storms. 91

Fig. 28, for next morning, November 18, still exhibits the traces of the cyclonic disturbance over Denmark, but the lowest readings there are about 29·7 inches, showing a rise of half an inch in pressure. All over the chart, except just in the south-west, the course of the isobars is changed and the advancing depression is per-

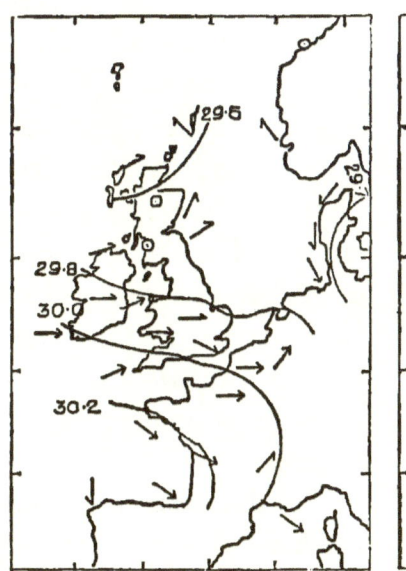 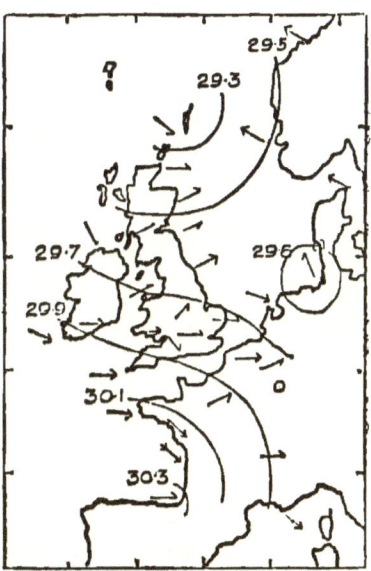

FIG. 28.—November 18, 1874; 8 a.m. New cyclonic disturbance appearing in north.

FIG. 29.—November 19, 1874; 8 a.m. Depression over Heligoland. A third appearing in the north.

fectly visible, and has produced its effect on the winds, causing them all to 'back' to W. or [SW. and even to SE. in Shetland. The morning of the 19th (fig. 29) shows us a great change; the [area of low pressure, which lay off Shetland on the previous day, has advanced rapidly southwards and lies over the mouth of the Elbe, while another similar system has appeared over Shetland. Each of these independent disturbances has its own atmospheric circulation.

The last chart of the series is fig. 30, for the 20th. In it we see the last-named depression situated in nearly the exact locality of its predecessor, over northern Germany, while the disturbance of the winds at the northern stations shows that the perturbations are not yet at an end. The rise, however, of the barometer over Holland indicates that the general distribution of pressure is undergoing a change.

In this case therefore we have a succession of cyclonic disturbances chasing each other for several days, along similar paths, while the region of highest pressure lies in the west.

If the highest readings be over France, the path of the cyclonic disturbances will run from west to east, across the British Isles, and the wind will blow hardest from the Westward. When the centre of the anticyclone lies over England we find that the storms hardly affect

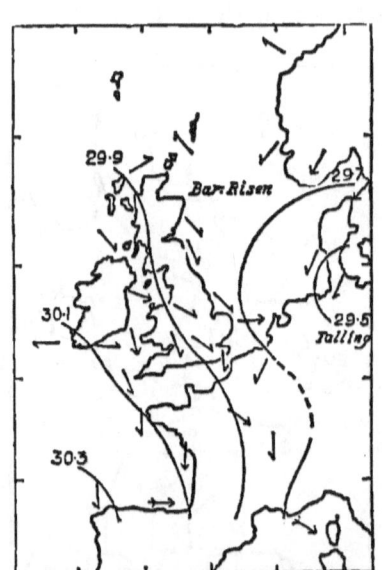

FIG. 30.—November 20, 1874; 8 a.m. Depression disappearing over Denmark.

The Motion of Storms.

our stations at all, but pass outside the north coast of Scotland, producing strong Westerly gales on the Norwegian coast, above the sixtieth parallel, at Christiansund, Bodö, and northwards. It is needless to cite instances of these conditions, they are already admirably illustrated by figs. 25 and 26, p. 87.

When the area of high pressure is located more to the east or south-east, over Denmark or Germany, the cyclonic systems cannot make good their footing over these islands at all, but pass from south to north outside the Irish and Scotch coasts, producing gales from the Southward. Such storms frequently only appear in the west of Ireland, and are thence propagated to the Hebrides, Orkneys, and Shetlands, without extending to the English coasts.

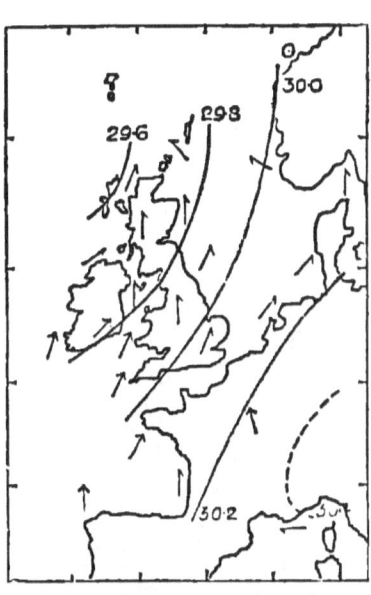

FIG. 31.—January 14, 1875; 8 a.m. Pressure highest in south-east. Depression off Hebrides.

A very good instance of this type of weather occurred in the beginning of the year 1875. Fig. 31, for January 14, shows us the barometer highest over Switzerland, and the isobars running NE. and SW., across these islands, while a depression is just visible off the NW. coast of Scotland. Fig. 32, for the 15th, shows but little change of pressure over central Europe, but a fresh disturbance has appeared off the west coast of Ireland, causing the winds to back to S. and SE., and blow with the force of a gale. Fig. 33, for the 16th, shows us this second disturbance in the position of its

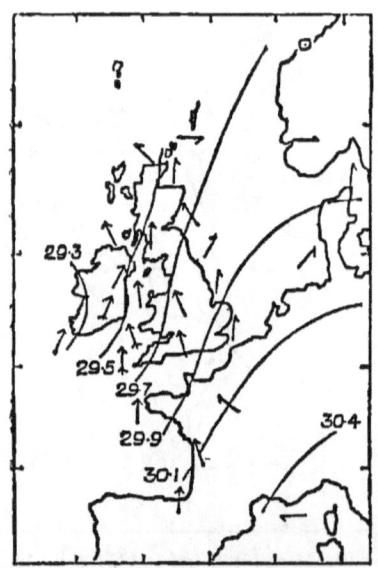 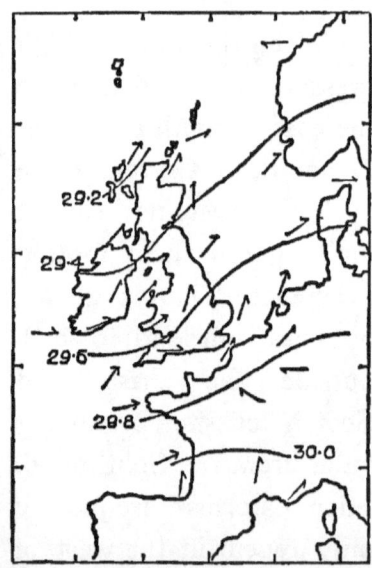

FIG. 32.—January 15, 1875; 8 a.m. New disturbance off west coast of Ireland.

FIG. 33.—January 16, 1875; 8 a.m. Pressure highest in south. Depression off NW. coast of Scotland, moving eastward.

predecessor of the 14th, but pressure having given way over Norway the conditions are altered, and the region of highest barometrical readings is transferred to Spain.

I have hitherto spoken of the cases when pressure is highest in the west, south, and east respectively, but the anticyclone must sometimes lie to the northward of

The Motion of Storms. 95

us, and then, if ever, should the disturbances advance from the eastward. Such a movement is, however, excessively rare in these latitudes, though this is not the case with the tropical hurricanes or cyclones, which at first move from the Eastward. There are, however, some principles which have not yet been thoroughly explained, and which are antagonistic to the development of such a motion in the storms of our part of the Temperate zone. I shall hereafter say a few words as to the causes which have been adduced to account for the motion of storms, but it will suffice, at this juncture, to remark that the motion of cyclones round anticyclones will not by any means account for all the motions which have been noticed in our storms.

The typical cyclone of November 29, 1874, figs. 19-22, pp. 83-85, to which reference has frequently been made, did not skirt round the region of high pressure on our charts, but travelled directly towards it. In this case, however, as in many others, it is possible that a study of the weather over more extensive charts, like those of Captain Hoffmeyer, might throw more light on this question of motion than we are at present able to obtain.

While discussing the motion of storms, it may be interesting to trace the path of a very erratic disturbance which visited these islands in April 1872, and the course of which has been followed out with the aid of the continuous records at our self-recording observatories. The following seven charts (figs. 34-40) show the successive

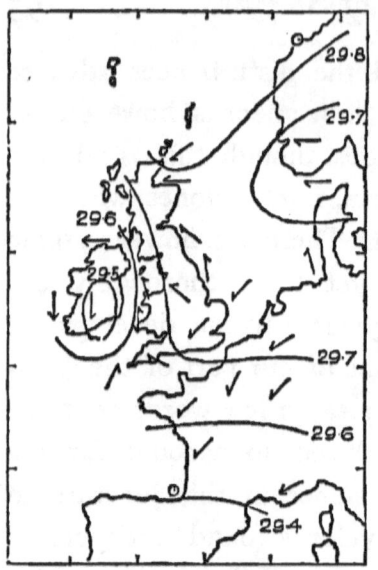

FIG. 34.—April 20, 1872; 8 a.m.
Centre of disturbance near
Waterford.

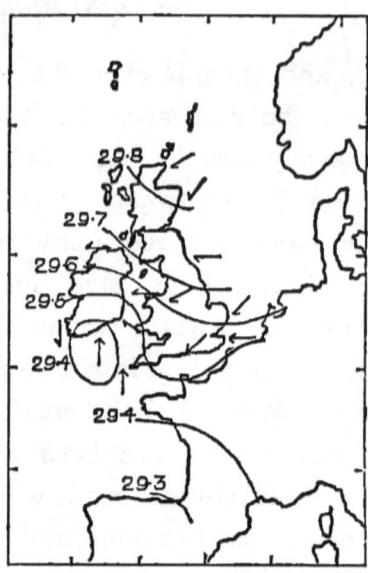

FIG. 35.—April 20, 1872; 6 p.m.
Centre of disturbance off south coast
of Ireland.

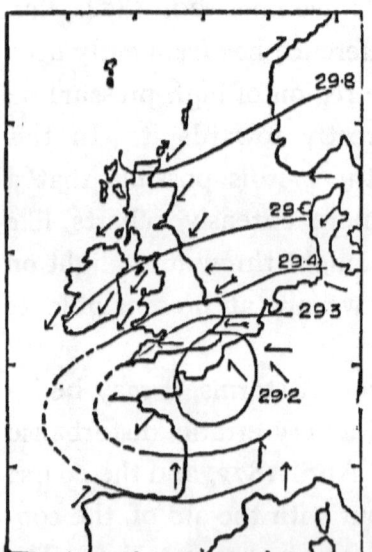

FIG. 36.—April 21, 1872; 8 a.m.
Centre of disturbance near Havre.

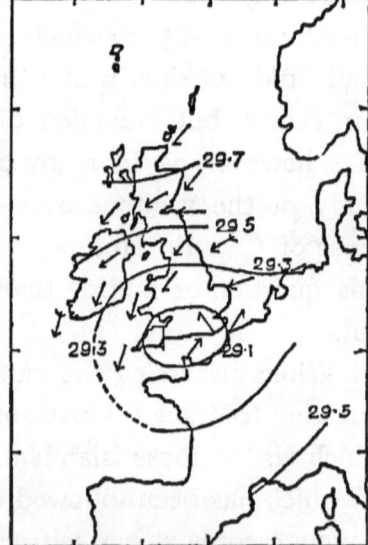

FIG. 37.—April 21, 1872; 6 p.m.
Centre of disturbance near
Portsmouth.

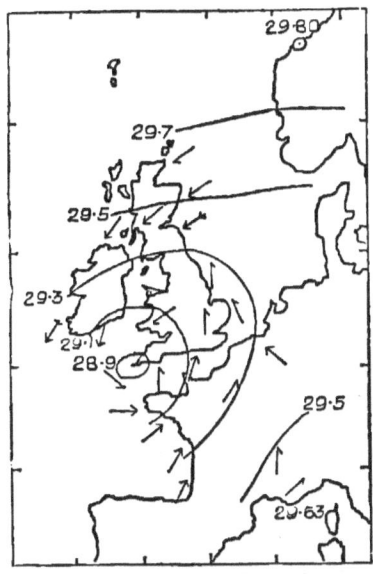
FIG. 38.—April 22, 1872; 8 a.m.
Centre of disturbance near
Penzance.

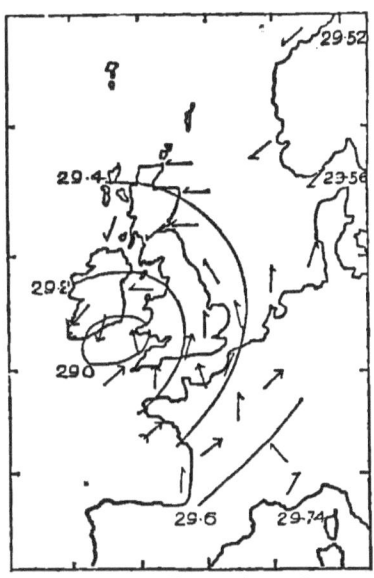
FIG. 39.—April 23, 1872; 8 a.m.
Centre of disturbance off south coast
of Ireland.

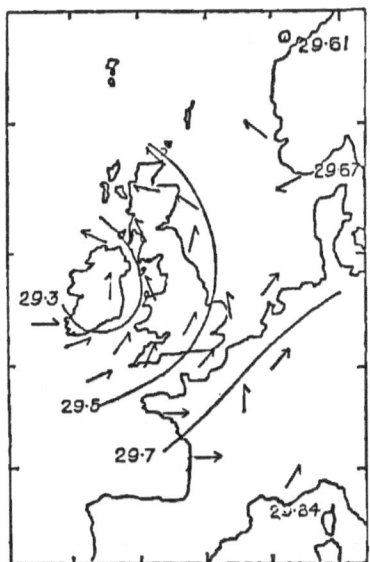
FIG. 40.—April 24, 1872; 8 a.m.
Disturbance passing off over Ireland.

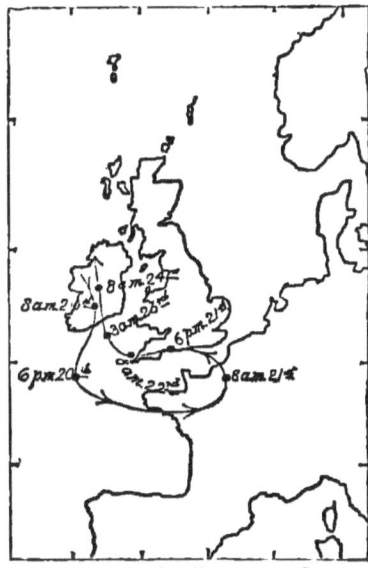
FIG. 41.—April 20–24, 1872.
Path of disturbance.

positions of the central disturbance, and fig. 41 gives its track for the entire period. It came in over Ireland and retreated along its own path again. That it did not advance as far as the Strait of Dover is clearly shown by the records at Kew, and that it passed first on one side and then on the other of Falmouth, is indicated by the records at that observatory, which prove by the direction of the shifts of wind, on the same principles as are explained in Chapter V., that the path of the centre lay first on the southern and then on the northern side of the station.

This storm presents us also with the rare phenomenon of an advance from the eastward. As far as we can at present form an opinion, this latter condition depends on circumstances of pressure far outside the area of these islands or even of that embraced by our weather charts. It is needless to say that such a storm as that just described afforded a striking instance of failure of warnings, as will be explained in Chapter VIII., p. 144.

The storms of which I have been treating have all exhibited motion more or less rapid, and in various directions, but it sometimes happens that areas of low pressure are stationary for two or three days together, nearly to the same extent as the anticyclone already noticed (figs. 23–26, p. 86). It is a remarkable fact as regards these islands that there are certain localities which apparently exert an attraction on these systems, and so retard their motion for a time. This is most strikingly the case with the region situated at the entrance of the

The Motion of Storms. 99

Channel, where, especially in the early autumn, cyclonic disturbances appear to remain almost immovable for days together.

A very good instance of a stationary depression occurred at the end of November 1872. Fig. 42, for

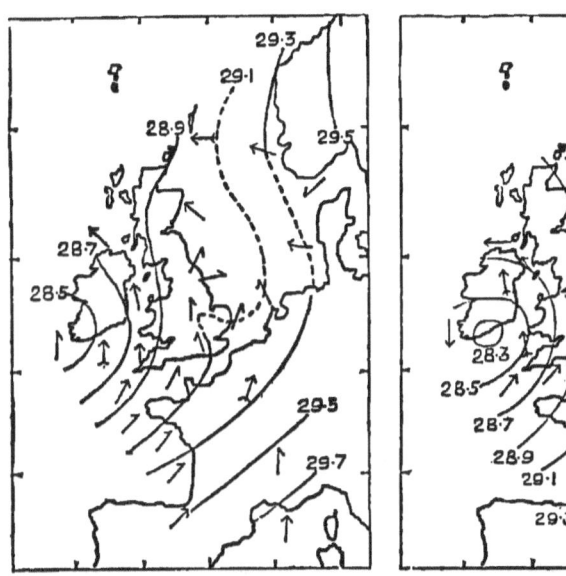 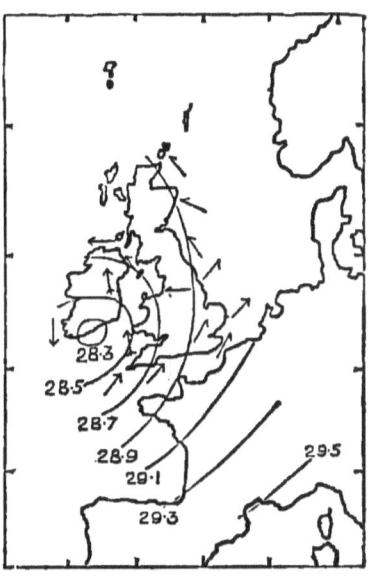

FIG. 42.—November 30, 1872; 8 a.m. Depression appearing off Valencia.

FIG. 43.—November 30, 1872; 6 p.m. Centre of depression near Cork.

November 30, at 8 a.m., shows us the isobar of 28·5 inches over the south-west of Ireland. There is nothing else very remarkable, except that readings are generally very low, and that a subsidiary depression appears off the coast of Lincolnshire. Fig. 43, for 6 p.m. on the same day, exhibits the centre of a storm over Cork, but matters otherwise do not show much change, except that pressure has given way considerably over France.[1]

Fig. 44, however, for 8 a.m. December 1, shows us the centre of the cyclonic area in nearly identically

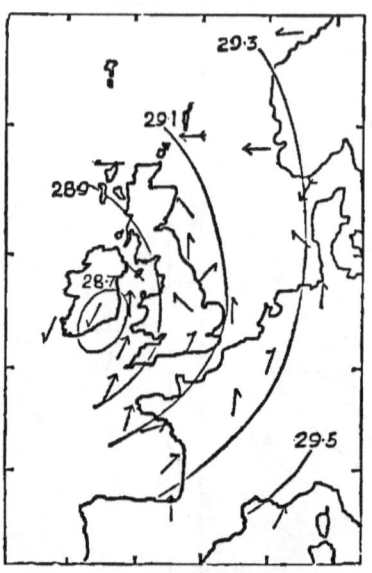

FIG. 44.—December 1, 1872 ; 8 a.m. Centre of depression nearly in same place, but depression filling up.

the same spot as it had been fourteen hours previously, but the whole system is being filled up, for the lowest isobar is that of 28·7 inches, while 28·3 inches was the minimum in fig. 43.

This disturbance subsequently took a course to the SSE., and passed on to the west of France.

[1] In 1872 we did not receive 6 p.m. reports from Norway, so that portion of the chart is blank.

In this chapter therefore I have shown that atmospherical disturbances advance over the earth's surface, and that cyclonic systems move more rapidly than anticyclonic. It has also been remarked that we know very little about the rate at which a storm is moving until it has advanced well within the area of our telegraphic reporting system. We are also virtually ignorant of what the real cause of this motion of storms is, whether it is due to the change of position of the entire mass of disturbed air, or to the production of a depression in front of the storm while it is filling up in the rear. We have traced some cases of motion of cyclones, and have stated that this is modified by the contour of the country over which they have to pass, but that it appears to be regulated in great measure by the position of the anticyclonic areas in the neighbourhood.

CHAPTER VII.

THE USE OF WEATHER CHARTS.

IF we come to consider what is the practical use, to an ordinary observer, of weather charts, giving, as they do, a representation of past conditions of weather, it will be necessary to recapitulate briefly the results to which we have been led in the several preceding chapters, and then to mention some of the most general principles on which a judgment is formed as to coming weather, when the observer knows by the study of a chart what conditions have been prevailing elsewhere. Subsequently I shall allude to some of the theories which have been propounded to account for the origin and motion of storms, and lastly, to the ideas which are entertained by some meteorologists as to a periodicity of rain and of storms, and as to a connection between disturbances in the gaseous envelope of the sun, and those which take place in our own atmosphere.

It will be useful in the first place to summarise briefly the principles which it has been endeavoured to establish in the preceding chapters.

Chapter I. has been devoted to an account of the

materials, in the way of observations, with which we have to construct our edifice.

In Chapter II. the wind has been discussed, as the atmospherical phenomenon which is most directly related to weather. It has been shown that wind is always connected with some disturbances of the pressure of the atmosphere, its existence of course being due to the tendency of an elastic body, like the air, to regain the condition of equilibrium from whence it has, by any means, been disturbed, while its motion is regulated by certain fixed laws.

The subject of Chapter III. is the barometer, as being the instrument by means of which the amount of atmospherical pressure, and therefore of its disturbances from time to time, is measured. It is shown how the distribution of pressure on a weather map is indicated by the isobars, and how there are two grand types of disturbance of atmospheric equilibrium, named respectively cyclonic and anticyclonic, according as the pressure is either in defect or in excess of its average value for the special chart in question. These two types of disturbance are characterised by different features, among which the most striking is the contrast in the direction of motion of the wind.

In Chapter IV. gradients have been explained, and it has been shown that the distribution of pressure, as measured by the gradient, is the best guide we have towards a knowledge of the laws of wind motion, and therefore, in some degree, towards a knowledge of coming weather.

It has, however, been stated that there are apparently other agencies than the distribution of pressure which influence the force of the wind, but that their precise nature has not yet been ascertained. It has also been pointed out that the phrases 'gradients for such and such winds,' &c., are nothing more than the expression of the laws of wind motion in a practical form.

Chapter V. contains a closer examination into the principal features of cyclonic and anticyclonic systems respectively, and into the character of the weather which accompanies each. It states that while anticyclonic areas are usually stationary or move slowly, cyclonic areas travel more or less rapidly over the country, and shows that their passage past a station presents marked features of difference according as that station lies on the right-hand or left-hand side of the path of the centre. This has been illustrated by the continuous automatic records of some of the self-recording observatories in connection with the Meteorological Office, and it has been shown how the usual changes of wind are such as are referred to in Dove's well-known Law of Gyration.

It has also been explained that cyclonic systems are not always isolated, but have occasionally what may be called satellite systems in connection with them, which exert a material influence on the gradients, and consequently on the winds belonging to the original disturbance.

Finally, Chapter VI. has been devoted to the motion of atmospherical disturbances, where it has been explained at more length than in the preceding chapter, that cyclonic systems move more rapidly than anticyclonic.

The Use of Weather Charts. 105

Some cases of the motion of storms have been traced, and it has been shown that while we know very little of the rate of motion of a storm before it arrives, we can frequently form a conception of the probable direction of motion of the storm. This motion is affected by the contour of the country over which it passes, and it appears in great measure to be regulated by the position of the anticyclonic areas in the neighbourhood.

The least consideration of the principles just enumerated will abundantly show that the weather prevailing over any district on a given day is clearly affected by, if it may not be described as the result of, the conditions prevailing in the districts round about, so that by observing what those conditions have been for some time back, we are enabled to form some opinion as to the weather which we are experiencing, and which is related to those conditions, whether it is likely to be permanent or transient in its character, and if transient, in what way it is likely to change.

It is hardly necessary to say that the fresher the information is the better, but owing to the cost, it is not possible in this country to follow the example of the United States, and send out by telegraph, in all directions, the entire mass of figures in the Daily Weather Reports, in order that it may be published simultaneously in every large town. Practically, in these islands the readers of newspapers and the subscribers to the Daily Weather Report cannot see their charts until several hours, frequently even an entire day, after date.

Such information, however, can generally be turned

to very good service, if rightly used, and I shall take the case of an observer resident not far from London, and see what he can do with it.

I have already explained that most of our disturbances travel from the westward, and as they take a certain time to advance over the distance which separates the east of England from the west of Ireland, we can learn, by studying the chart for the previous day, what conditions were then existing at the western stations, and consequently we can gain some idea of what is likely to be the result of the changes which we notice by our own local observations, instrumental and otherwise.

If, for instance, the barometer at our station is high and steady, with dry weather and light winds, either in winter or summer, we may form a general notion that the type of weather is anticyclonic, and as this type is peculiarly permanent, we may feel sure that any change will give us at least several hours' notice of its coming, by alterations in instrumental readings occurring over some part of the district covered by the Reports. The study of the charts in this case is the more important, because frequently, at such a time, 'cirrus' or 'mare's tail' clouds appear in the sky, which are usually the precursors of coming wind; and change of wind in such a case means change of weather. The charts then will show whether or not this wind, existing at a great elevation, had made its way down to the earth's surface at any place on the previous day within the district just referred to.

The Use of Weather Charts. 107

In the case of cyclonic conditions, we have already learnt that it will not do to trust implicitly to the fact of storms advancing from the westward, but on the other hand some ideas have been thrown out by means of which we can, at least in some cases, forecast the probable direction of motion of a storm. At any rate we are usually able to learn from a careful study of the chart, whether or not we have to deal with disturbances on a large or a small scale, and accordingly, whether the changes in progress indicate lasting, or merely temporary, conditions of weather.

These charts are therefore useful helps to the local observer, and will be found so by those who *study them regularly, and combine with that study careful and systematic observations of their own instruments, and of local weather;* but for the purpose of judging whether a particular afternoon will be wet or fine, which is all that the public generally care to know about weather, it is obvious that charts which are in many places necessarily twenty-four hours old cannot be of much service. Moreover the phenomena which we include under the general term 'weather,' often depend in great measure on the nature and conformation of the ground in the neighbourhood of the observer, so that one place will be much more liable to rain during disturbed weather than another, while a second will exhibit a greater tendency to the formation of fog at a calm period than an adjacent district might show. As, therefore, such exceptional tendency is confined to each special locality, and does not belong to the phenomena produced at all stations by the system of circulation prevailing at the time, it is necessary that the observer who endeavours to forecast probable weather should seek to ascertain under what conditions such peculiarities manifest themselves, as it would be useless to apply merely general rules in order to see the meaning of phenomena of a purely local character.

On the whole it must be said that our insular and exposed position precludes us, in the present state of our knowledge, from the possibility of issuing forecasts of future weather sufficiently trustworthy to be worth publication, excepting occasionally, and then principally for the south-east of England.

Let us now examine in what manner the knowledge we have gained would enable us to issue storm warnings to the coasts. In the first place we can in many instances learn from the general conditions of wind, &c., in fact of weather as a whole, what will be the character of the changes likely to occur, and in what direction they will propagate themselves.

We then notice the conditions of pressure as shown by the barometer, what its changes are, and whether any changes which may manifest themselves are traceable at adjacent stations, and in what degree, so that we can gain some idea of the extent of the disturbances which are approaching.

We next see in what degree the barometrical changes are borne out by the wind, both in direction and force; in fact we often get earlier intimation of approaching change of weather from abnormal features in the direction of the wind, than from the barometer readings, taken by themselves.

Temperature, too, must be looked to, for as a general rule, in winter, if the temperature has been very low, even though the wind has been Southerly, we need not fear a Southerly gale until the advancing depression has exerted a marked influence in raising the temperature.

We have then the reports as to weather at the different stations, and the deficiency of information of this nature has been amply explained in Chapter I. To this class of observations belong reports of the character and motion of clouds, and of the clearness or mistiness of the air. Each of these latter may be, in their way, a prognostic of storm; the former, an unnatural clearness, being a very bad indication before the storm has actually reached us and affected our winds, while the latter, mistiness, is often the immediate forerunner of coming rain, and sets in when the wind in an advancing cyclonic system has begun to blow from the South or South-east, or more especially on the ridge of high pressure or debateable ground, which lies between the NW. wind of a *retreating*, and the SE. wind of an *advancing*, area of low pressure.

Reports of the Aurora are also very useful as corroborative evidence. The past winter (1875–6), since the beginning of December, has been the freest of any for the last nine years from Westerly winds, *i.e.* from cyclonic disturbances passing to the northward of these islands, and it has also been remarkable for the rarity of reports of the Aurora.

It is, however, obvious that we, in the Meteorological Office, labour under prodigious disadvantages in attempting to issue warnings for the United Kingdom. Interest therefore attaches itself to the attempts which we are making from time to time, to draw conclusions from the experience of past weather which may be useful to us in forming opinions on what is to come.

Attention has already been drawn to cyclonic disturbances which follow each other along the same track as long as the general distribution of pressure remains unchanged, and it has also been stated at p. 15 that cyclonic disturbances travel across the Atlantic for long distances, if not from shore to shore. This is abundantly proved by the well known fact that steamers, when outward-bound to America, often meet a succession of such cyclonic systems on their way towards Europe, and when homeward-bound to England, often run for a certain time with such a system, so that in the former case the changes in the instrumental indications, as well as in winds and weather, are much more rapid than in the latter, on the simple principle that if walking in a crowded thoroughfare you meet many more vehicles than you pass, or than overtake you. I shall return to this subject of the advance of storms from America at p. 126.

If, then, these disturbances are frequently travelling in succession over the Atlantic, it is evident that we in these islands must at times encounter a series which bears a close resemblance to a series which has preceded it as to the general character of the areas of depression themselves, and as to the intervals which occur between them. This we have noticed in our own weather on various occasions, and the parallelism has been found to hold good for as many as eight days together. Here, then, is a most promising field for enquiry, which it is hoped will eventually be carried out.

As to the theories which have been of late years propounded to account for the origin and motion of storms, I shall not attempt to discuss them, but merely indicate the broad features of some of them in a very cursory way.

Professor Mohn, the Rev. W. Clement Ley, and others, attribute the generation and motion of storms to the condensation of moisture in the form of clouds, resulting in rain. This condensation takes place principally in front of the storm, and, so to speak, draws it on.

Mr. Thomas Belt and Professor Reye account for the origin of the storm by the existence of a condition of unstable equilibrium in the atmosphere; a cold stratum being situated at some height above the ground, if the air at the earth's surface becomes heated, it must eventually force its way through the superincumbent colder layer, and the upward current thus generated will be the core of the resulting cyclone. Professor Reye goes so far as to say that the storms of the largest dimensions experienced in these latitudes are substantially analogous to the smallest waterspouts or dust-whirls we can observe.

In direct contradiction to this view, M. Faye, in the 'Annuaire du Bureau des Longitudes' for 1875, has propounded the view that cyclones are vortices descending from the upper regions of the atmosphere to the earth's surface, and that their motion is that of the upper current.

Lastly, Mr. Meldrum and others consider that cyclones are always generated in the space intervening between two currents, which themselves are tangents to the nascent cyclone. Thus, in the North Temperate zone, the conditions which would give rise to the cyclones are those which will be noticed at p. 131, of Easterly winds on the northern side of Westerly winds. We need hardly remind our readers that every perfectly developed cyclonic disturbance here must have East winds in the north and West winds in the south.

It is clear that when 'doctors differ' to the extent indicated in the above sketches of the different theories, the world in general must be content to wait patiently for a complete theory of weather to be developed.

Before we leave the subject altogether, it will perhaps be interesting to see the extent to which we are able to forecast the weather for longer periods than a day or two. Attempts have not unfrequently been made to predict the seasons for a long period in advance, but without much success hitherto. One great cause for failure is that accurate meteorological records do not extend beyond the beginning of the present century at more than a few stations, and at these we are unable to eliminate the local influences altogether. Thus, it is

The Use of Weather Charts. 113

hardly possible to say what has been the approximate temperature of these islands for more than twenty years—a period far too short for the definite recognition of a cycle. The shortest of such cosmical cycles which has been determined is the sun-spot period of $11\frac{1}{9}$ years, according to Wolf, and there are indications of far longer periods, such as 33 years, or even $69\frac{1}{3}$ years, according to Hornstein.

Of late years Mr. Meldrum, of the Mauritius, has shown that the cyclones for which that district of the Indian Ocean enjoys an unenviable notoriety, have been more frequent in some years than in others, and that these epochs of maximum frequency occur at intervals of about eleven years, coinciding with those of maximum sun-spot frequency.

This agreement is most important, and it appears to be corroborated by an examination of the rainfall at several stations which has been conducted by Mr. Meldrum and others. The results, for the comparatively short period to which they refer, are very striking, and are sufficient to show that a periodicity is traceable in the weather of the Southern Indian Ocean, which is eminently suggestive of an intimate relationship between the changes which take place on the sun's surface, and the phenomena of our own atmosphere.

It will at once be asked, Why has not this periodicity, if it exists, been detected long ago by an examination of European records, which are far more complete than any existing for the Indian Ocean? The answer to this is twofold.

I

In the first place we are pre-eminently in the region of the variable winds, and our storms are not nearly so regular in their type as those of the Mauritius, where almost the sole type of storm is the true tropical cyclone with its concomitant rainfall. It is next to impossible in this country to keep a record of all the storms which pass over us. We have already seen that the existence of conjugate storms is not unfrequent: two or even three systems of disturbance being traceable at one time within the limited area of the United Kingdom. Are these one single storm or several, and how should they be counted in a catalogue? Rain also cannot be taken as a sign of the frequency of wind storms in a year; for although we know that warm winters are invariably wet and stormy, and moreover, that cyclonic disturbances are accompanied by rain at all seasons, yet it cannot be asserted that either the almost constant rain of 1872, or the recurrence of floods in 1875, were in any way related to storms of wind, though their connection with the presence of areas of depression over these islands is indisputable.

There is, however, in the second place, a far deeper reason for the non-discovery of these cycles in any chance series of rainfall records. The sun passes through phases of greater and less activity, and the terrestrial phenomena corresponding to the epochs of the former character are excessive evaporation in some parts of the globe, and consequent excessive precipitation in others. We must therefore ascertain in what districts we are to

The Use of Weather Charts.

look for the one and for the other of these phenomena respectively. In fact we cannot yet say where we shall find the maximum solar effect produced.

The lesson that we are to learn from the fruitful researches of Mr. Meldrum has been pointed out by himself and others ; it is that we should aim at attaining a thorough knowledge of the movements and changes of our own atmosphere, and should then seek to establish a connection between them and other cosmical phenomena, such as terrestrial magnetism, the relation of which to the state of the sun's surface was pointed out by Sir E. Sabine more than twenty years ago.

CHAPTER VIII.

STORM WARNINGS.

STORM Warnings may be considered as the most immediate practical application of weather knowledge, and in fact it was the possibility of issuing such intelligence which gave the first impetus to the development of Weather Telegraphy. Although this branch of the subject has not yet been reduced to such strict rules of procedure as are demanded by the requirements of an exact science like Astronomy, meteorologists can at all events lay claim to having established some principles which are sufficient, in many cases, to throw valuable light on the conditions which may be expected to result from existing circumstances, and so to some extent to lift the veil which shrouds the future from our eyes.

The first suggestion of the electric telegraph as a method of conveying intelligence of storms from one place to another is apparently due to Mr. Redfield in America, who shares with our own countryman, Sir W. Reid, the honour of having reduced the Law of Storms to a science. Redfield's paper is in the 'American Journal of Science and Art' for 1847.

Storm Warnings.

The subject was taken up in a tentative way in the United States between the years 1850 and 1860, but further operations in this direction were brought to a standstill by the war in 1861.

On this side of the Atlantic the credit of taking the initiative is due to LeVerrier, the present director of the Paris Observatory, who, as early as on February 17, 1855, received the Emperor's sanction for the creation of an extensive organisation destined to distribute intelligence of weather over the countries embraced by it. The possibility of rendering such an organisation really useful had been previously shown by the study of the Balaclava storm of November 14, 1854, which had wrought immense damage to the allied fleets in the Black Sea, and by whose effects hardly a country in southern Europe had been left unscathed. It was evident, from a mere cursory study of the facts of this storm, especially of its rate of progress, that timely notice of its approach might have been issued from the coasts which were first visited to those which subsequently felt its effects.

The plan gradually assumed a definite form, and in the spring of 1860 M. LeVerrier addressed a letter to Sir George Airy, inviting the co-operation of this country in his scheme. This letter contains certain expressions which have such an important bearing on the subsequent history of storm warnings that they are worth translation and quotation. They are as follows:

'The ultimate result of the organisation which we

are establishing should be to announce a storm as soon as it appears at any point in Europe, to follow it on its course by means of the telegraph, and to give timely notice of it to the coasts which it may reach.'

I shall proceed to show to what extent this programme has been carried out, and to what extent it has been possible to go beyond it.

In this country the idea of storm warnings had been broached before 1860, for at the meeting of the British Association at Aberdeen in 1859, a formal resolution was adopted in favour of the organisation of such a system. A month after the meeting the loss of the 'Royal Charter' on the coast of Anglesea arrested universal attention, and as there seemed, at first sight, a chance that that ship might possibly have been warned of her danger, the hasty conclusion was drawn that no storm could ever come on without giving timely indication of its own approach. Subsequent experience soon proved how very unsafe such a generalisation was.

Early in 1861 the first tentative warnings were sent out, and by the beginning of 1862 Admiral FitzRoy's system was definitely established. The principles on which our present warning system is based are mainly the same as the principles of that devised by the Admiral, but exhibit some contrasts to them.

Admiral FitzRoy, when he issued a signal, intended by it to imply that the storm of which it gave warning would be likely to occur within the next seventy-two hours. He therefore had the signals kept up until the evening of the day on which they were hoisted, and then lowered, while by the present system they are not lowered until all danger of the gale appears to be past. It is evident from this that if, as often happens, threatening symptoms disappeared without the actual occurrence of a gale, Admiral FitzRoy had no means of announcing to the ports, on the day after the issue of a warning, the fact of the improvement in the weather prospects. This arose from the fact that once his signal had been lowered there was no public indication that it had even been hoisted at all, and yet the warning had been intended to cover three whole days. By the present system the fact that no signal is exhibited is, to a certain extent, a proof that no danger is apprehended at the Office in London.

It is hardly necessary to describe the signals, as their appearance is familiar to all who have been at seaside places, but for the sake of those who may not have been on the coast in winter, at which season the signals are most frequently visible, I may give the following explanation of their meaning.

TELEGRAPHIC WEATHER INTELLIGENCE.

The fact that a warning has been received at any station is made known by a signal, which is hoisted on the receipt of the message, and remains hoisted, but only during the day-time, for the space of forty-eight hours and no longer, counted from the time the message is sent out.

The signals are made by means of two canvas shapes, a *cone* and a *drum*, fig. 45, p. 121.

The *cone* is three feet high and three feet wide at base, and appears as a triangle when hoisted.

The *drum* (or cylinder) is three feet high and three feet wide, and appears as a square when hoisted.

The *cone, point downwards*, means that strong winds are probable, at first from the Southward (from SE. round by S. to NW.).

The *cone, point upwards*, means that strong winds are probable, at first from the Northward (from NW. round by N. to SE.).

The *drum* is hoisted with the cone whenever an unusually heavy gale, either Southerly or Northerly, as the case may be, is probable.

The drum is not used without the cone.

It must be remembered that a Southerly wind is much more likely to veer suddenly to a point North of West than a Northerly wind is to veer to a point South of East.

Accordingly, when the South cone is hoisted, and the anchorage or harbour is exposed to the North-west, it is advisable to make preparations for a North-west gale.

The signal is kept flying until dusk, and then lowered, and hoisted again next morning; and so on until the end of forty-eight hours from the time the message has been issued from London (which is always marked on the telegram), unless orders are received previously to lower the signal.

Storm Warnings.

CAUTIONARY SIGNALS.

Day signals.

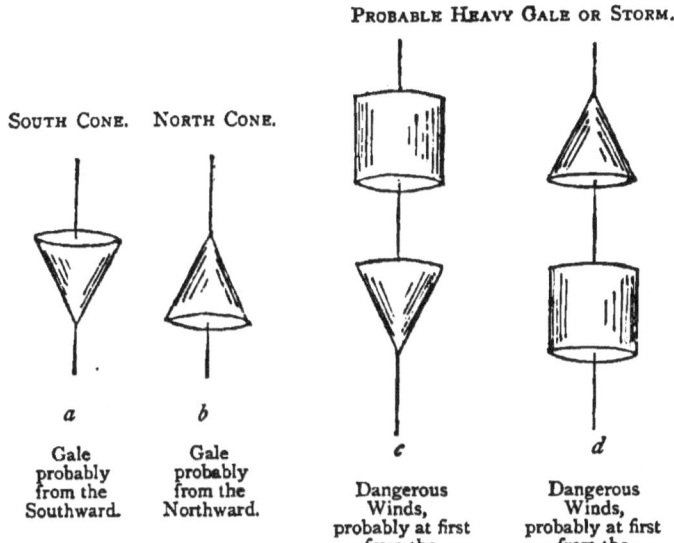

Night signals (instead of the above), lights in triangle.

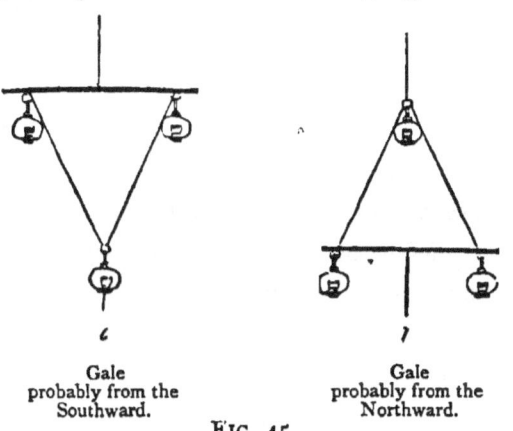

FIG. 45.

At dusk, whenever a signal ought to be flying if it were daylight, a night signal may be hoisted in place of the cone, consisting of three lanterns hung on a triangular frame, point downwards or point upwards, as the case may be. It is not considered necessary to hoist lanterns to represent the drum. The lanterns are kept burning until late in the evening, say nine or ten o'clock.

MEANING OF SIGNAL.

The hoisting of any of these signals is intended as a sign that there is an atmospherical disturbance in existence which will *probably* cause a gale, from the quarter indicated by the signal used, in the neighbourhood (say, within a distance of fifty miles) of the place where the signal is hoisted, and the knowledge of which is likely to be of use to the sailors and fishermen on that part of the coast. Its meaning is simply, 'Look out! It is probable that bad weather of such and such a character is approaching you.'

In every case some of the principal reasons which have led to the hoisting of the signal are explained in the telegram, *which should always be kept posted up for public inspection while the signal is flying.*

The simplicity of the idea of using a cone and a drum, which must necessarily show the same shape, as a triangle or a square, no matter how they are blown about by the wind, is the great merit of FitzRoy's plan, as the sailor need not burthen his memory with a complex code of signals. In fact when the Meteorological Office in 1868 made an attempt to introduce a more complex apparatus, like a railway semaphore, in order to convey intelligence to ships in the offing, the experiment fell perfectly flat, owing to the difficulty of introducing a totally novel system for the use of our coasting seamen, for whose benefit storm warnings are originally intended.

The question of signals is not an easy one to set finally at rest. There is no doubt that the information conveyed by them would be infinitely more valuable if they could be shown from headlands which shipping usually make, such as the Lizard, than if they were simply exhibited in harbours. Here then we see the necessity for a special signal code. By our present system we can only give a limited number of possible signals, so that we cannot convey detailed information to ships at any distance from the shore; say, even riding at anchor in the Downs.

Our system is to hoist the signal as an indication that some information of a certain kind has been sent to the signal station, which it is of importance to the seaman to know: but he must land and read the telegram himself if he wants to learn more precisely what the facts are. We are as yet a long way off from the condition which may be described as the acme of weather-signalling, and which has often been suggested, that of having a weather signal book containing charts of different types of weather, much like the illustrations in these pages, duly numbered, so that by simply hoisting a number on the signal staff, we could tell the seaman to what page in the book he had to turn, and what weather he had to expect. Thus such a chart as that for November 29, 1874 (fig. 2), might be described as No. 371, and so on. It will, however, be many a long year before our knowledge of weather will justify our issuing such a collection of typical weather maps as would be likely to be of service for such a purpose, but we may fairly cherish the hope that in the future something of this kind may be done.

It may here be observed that the Meteorological Office is often expected to be far more venturesome in its announcements of weather than it is at all prepared to be. Not very long ago, a gentleman called on me with the request that I would prepare for a newspaper which he had recently started, a forecast of the probable weather for a week in advance. On my declining the task, he remarked that Admiral FitzRoy had issued forecasts for three days in advance, and that by this time we ought to be able to do so for double that period. He quietly ignored the fact that Admiral FitzRoy's three-day forecasts were not successful enough for them to be continued, and that accordingly they have not been kept up.

Other men call on us to be as precise in our announcements of 'probabilities' as the Chief Signal Office at Washington, in simple oblivion of the circumstance that while that office has an immense continent to deal with, and is itself situated on the eastern side of its area of observation; so that the gradual march of weather changes over the western stations can be observed and reported long before they reach the more thickly populated portions of the States; we have to deal with an extended oceanic coast, and can by no possibility have a series of stations on our western side, so that, unless by the appearance of the clouds and sky, we have no chance of learning what is going on, out of sight of land.

Our reporting system is shown on the frontispiece, and a glance at it will prove how badly we are off for stations on the west coasts of Ireland and Scotland. These are the most important districts for gaining early intelligence of weather changes, but in the first instance telegraphic communication is only very scantily developed in such wild regions, and in the second place the habitable spots are all in sheltered bays, where the true force of the wind is hardly ever felt, owing to the fact that few ordinary houses could bear the brunt of a winter Atlantic gale, blowing, as it sometimes does, with a velocity of over 80 miles in a single hour, and reaching the speed of at least 100 or 120 in gusts. The observers therefore cannot possibly send up perfectly true reports of the direction and force of wind, as will be at once admitted when the position of the stations is considered. Valencia lies on the shore of a narrow sound, with high hills about it. At Greencastle the high land of Innishowen breaks the force of the South-west and West winds. Ardrossan is a valuable station, but it is far up the Frith of Clyde, and at least fifty miles, as the crow flies, from the extreme western point of Islay. We have no other station till we reach Stornoway in the Island of Lewis.

A great deal is talked about our getting reports from Newfoundland, the United States, the Azores, and the Faroes, and lastly from moored signal ships.

We received telegrams for a long time from Heart's Content *gratis*, through the liberality of the Anglo-American Telegraph Company, but we could not turn them to practical account, partly owing to the fact that the situation of that station was chosen as a sheltered nook, where no storm could hurt the cable, so that the wind felt there bore little relation to that blowing at sea outside; but more particularly on account of the circumstance that though storms do sometimes cross the Atlantic from shore to shore, they change their character *en route*, some increasing and others dying out, so that we cannot possibly predict which storm, out of several starting from the States, will reach us. Professor Loomis, a very high authority on American weather, supports the above view very strongly in the following words, taken from the 'American Journal of Science and Art' for January 1876.

'When storms from the American continent enter upon the Atlantic Ocean they generally undergo important changes in a few days, and are frequently merged in other storms which appear to originate over the ocean, so that we can seldom identify a storm in its course entirely across the Atlantic.'

To give an instance of a storm failing to advance over any great distance, I shall take one which will be reasonably fresh in the memory of all. On November

30, 1874, the steamer 'La Plata,' with a telegraph cable on board, foundered in a heavy gale near the Channel Islands. This storm entirely died out and disappeared before it crossed the North Sea.

These statements show how rash is the opinion which is held by Professor Daniel Draper, of New York, and announced in the Reports of the Central Park Observatory for 1872 and 1873, to the effect that out of eighty-six storms which he had investigated, as having started from New York, only three had failed to show themselves either at Valencia or Falmouth. This assertion has cropped up again and again in newspapers and magazines at this side of the Atlantic, but it is only fair to ourselves to say that the idea was received with even greater incredulity by meteorologists in the United States than it met with among scientific men over here.

To take a matter nearer home ; it has often been suggested that this country should bear a considerable share in the cost of daily telegrams from the Azores to Europe, as it was assumed that as the islands in question are situated in the Atlantic, off the coast of Portugal, they must be, so to speak, pickets to give us early news of danger approaching from that quarter.

All this is very plausible, but when we come to test the simultaneous reports taken daily in those islands and at Valencia, as has been done at the Meteorological Office for two years and a half ('Quarterly Journal of the Meteorological Society,' vol. i. p. 183), we find that there is practically no connection between the phenomena occurring at the two stations, so that we could get no *direct* warnings of approaching storms from the Azores, though of course a daily knowledge of the conditions of weather over the part of the Atlantic where those islands are situated could not fail to be of value to us, as is evident from what has been said in Chapter VI. In fact the Azores lie to the southward of the usual track of Atlantic storms, and it would apparently be a far better investment of money for this country to spend it in obtaining reports from the Faroes, if ever a cable were laid to those islands, than in paying for telegrams from the Azores, as many of the most destructive storms on our north-east coast come down on us from the northern part of the Atlantic Ocean and pass over the Faroe Islands. The value of St. Kilda as an outpost could hardly be overrated, but there is no present hope of telegraphic connection with that island, and in all cases it must be remembered that unless a cable will pay commercially it is hopeless to look for the establishment of telegraphic communication for scientific purposes.

The idea of the employment of ships moored at a distance from the coast, and connected by a telegraphic cable with the shore, has already been noticed in Chapter I. This plan has over and over again been suggested as a most valuable mode of obtaining information, not only of the position of vessels in distress or wind-bound, but of the meteorological conditions prevailing out at sea. Some scientific men, as Mr. Morse, have even gone so far as to propose that these signal ships need not even be manned, but might be simple buoys provided with electrically self-registering apparatus, which might record its indications automatically at the nearest station on land!

The plain state of the case about any such visionary projects is that, even if we were to overcome the difficulty of mooring a ship in 1,000 fathom water, we should find it next to impossible to maintain the telegraphic cable in perfect working condition. Some years ago the Admiralty lent H.M.S. 'Brisk' to be moored as a trial signal ship at the entrance of the Channel. The experiment was given up in six weeks, mainly on the ground of the difficulty of maintaining the telegraphic connection in a perfect state, and I believe that the venture cost the promoters about as many thousand pounds as the 'Brisk' was days at sea.

We learn therefore from all this, that we in the United Kingdom have to rely on our own resources and our own stations, for gaining information for storm-warning purposes, and so we must only do our best to improve our own arrangements.

It is evident that, if we had an extensive continent like North America to deal with, we too, in the Meteorological Office, could probably make good our claim to a proportion of successes for our warnings, rivalling that with which the Chief Signal Office is justly credited. That this is not an idle boast is shown by the following figures taken from the Report of the Meteorological Office (the Deutsche Seewarte) at Hamburg for 1874.

K

A system of warnings for Hamburg from our Office has been in operation since 1867, and the general results are, that 301 warning messages were issued from London in the course of the 7 years. 72 per cent. of these warnings were followed by gales, while *in only three cases* did the storm outrun the message. Naturally in these figures no account is taken of the gales missed owing to the Sunday interruption of our service.

The reason of this success in issuing to Hamburg timely warning of all storms, which is far more brilliant than we can ever show for our own coasts, is that our network of stations, as will be seen from the frontispiece, entirely surrounds the North Sea, and that no storm can reach Hamburg without passing over our outposts, unless it comes from the continental side, a very unusual direction for its advance. As a proof of this statement I may say that in the year 1869 twenty-three storms were felt in Hamburg, and twenty-two of these had previously passed over some part of the United Kingdom.

If it be asked what the real practice of storm warnings is at present, I can only reply that it is a moderate advance on LeVerrier's anticipations in 1860, p. 117. It will appear from the previous pages that we have some sort of knowledge of the circumstances under which the various types of storms approach the different parts of our coasts, but these are not sufficient to tell us in every case when a storm is imminent.

Some conditions, like those pointed out in a paper of mine printed in the 'Proceedings of the Royal Society' for 1869,[1] give us more warning than others. In that paper it is shown that the appearance of Easterly winds on the *northern* side of Westerly winds is a nearly sure sign of the approach of Southerly winds, the precursors of a cyclonic area, and perhaps of a storm, which will probably affect the whole kingdom, while the appearance of Easterly winds on the *southern* side of Westerly winds is not followed by any disturbance of a cyclonic character. In all this we are of course speaking of the northern hemisphere.

These circumstances, and other premonitions of a similar nature, are not of very frequent occurrence, and in general it is not until the storm is quite close at hand, and the barometer has begun to fall briskly at the outposts, that we feel ourselves justified in issuing a definite warning. The appearance of one of the small satellite depressions, however, not very unfrequently precedes that of a more serious storm; accordingly, if we notice one of these lesser disturbances, we are at once put on our guard.

A rapid and unexpected rise of the barometer is often the precursor of a coming depression, so that whenever we see a sudden rise we may expect an equally sudden fall, and must be on the look out for the slightest tendency of pressure to give way.

It should, however, be remembered that, as regards

---

[1] 'On the connexion between oppositely disposed currents of air, and the weather subsequently experienced in the British Islands:' Proceedings of the Royal Society, vol. xviii., p. 12.

barometrical indications, we are guided not only by the actual fall of the mercurial column on a certain line of our own coasts, but also by the behaviour of the barometer at far distant continental stations. Thus in the case of the storm already so often quoted, November 29, 1874, a warning had been issued to the coasts at 3.30 p.m. on the 28th, when the most alarming sign was not so much the fall of the barometer at our south-western stations, taken by itself, as the fact that this was accompanied by a slight rise, of 0·01 inch, since 8 a.m. at Rochefort, readings being already seven tenths of an inch higher in the south-west of France than in the south of Ireland. This showed that pressure was banking up to the southward, and that gradients were consequently becoming steeper along the Channel coast. If, when the barometer falls at our south-western stations, the fall affects those in the south of France, we know that we have less to fear from Westerly winds, as the tendency of such a change is to reduce the gradients for winds from that quarter, while a rise of pressure over France when it is already relatively high there, puts us at once on our guard for a blow from the Westward.

Similar reasoning will apply to barometrical indications in other parts, but of almost equal value to them, and in some respects of even higher importance, as giving earlier intimation, are the changes in direction of the wind.

I have already alluded to the proverbial danger of a backing wind p. 72; this, however, is chiefly true as concerns the 'backing' of the wind from NW. to SW. and S., which indicates the approach of a fresh disturbance, for it has been explained that if one cyclonic storm is following another in quick succession, as the wind on the southern side of each runs gradually from SE. through S. and W. to NW., the approach of the second depression will cause the vane to shift back from NW. towards SE. This backing is almost the most dangerous sign of all.

It has, however, been stated already that it is not safe to depend on the observations from our western stations as to the direction and force of the wind, owing to the hilly nature of the coasts, so that we cannot implicitly trust the reports on this subject which we receive.

Another very important sign as regards the wind, is the indraught of air towards the storm area. This is perfectly intelligible on the supposition that the air really flows in towards the centre and rises there. On the approach of a cyclone to the west coast of Ireland, we frequently find the winds South-easterly on the west coast of France, as on November 28, 1874, p. 83, while we can hardly say that the barometer at such a distance from the centre of the disturbance had yet felt its influence.

The reports of sea disturbance are even more uncertain guides to us than the observations of wind, as is explained in Chapter I., p. 11.

Temperature is also a very useful aid in gaining a knowledge of coming disturbance, for it will be remembered how the rise of the thermometer in front of a storm is a marked characteristic of areas of low pressure, but the indications of temperature have, for various reasons, not yet been reduced to strict rules like those of the barometer.

We know, however, that a great contrast of temperature between adjacent stations or, so to speak, a great thermometric gradient, being an indication of serious atmospherical disturbance, is the precursor or concomitant of a serious storm, so that if we find a great difference between the reports of temperature, we are at once warned of approaching disturbance. A recent instance of this fact may be cited. On November 13, 1875, at 8 a.m., the thermometric report from Scilly was 57°, and from Wick 21°. The resulting difference is 36°. The gale of Sunday, November 14, with its accompanying high tide and consequent inundations on the south coast and in London, will be fresh in the memory of all.

Besides all the symptoms of the approach of a storm which have been enumerated, there are the innumerable local signs which enable the fisherman and shepherd, not to speak of the inferior animals, to judge of coming bad weather. Among all these, almost the most important are the character and motion of the upper clouds, showing the existence of wind aloft which will, in all probability, soon descend to our level. Another great sign is the clearness of the atmosphere, the unusual visibility of distant objects, which is well known to all as a sign of a coming gale.

It must always be remembered, that in order to issue perfectly correct storm warnings, we should require to know the size, shape, position, and motion, in direction and rate, of an advancing depression, and also whether it is becoming deeper or the contrary, and that *there is not one of these conditions of which we have a really sufficient knowledge at present*, while of most of them we can have no knowledge at all till the storm has burst upon us. The problem which is put before the Meteorological Office daily, is similar to one which astronomers would at once recognise as impossible of solution, and that is to determine the elements of a comet's path from a single observation taken, say, in a brief clear interval on a cloudy night. The first glimpse we get of a storm must suffice for us to issue our warnings.

It is therefore evident that for our own exposed western and northern coasts we can but rarely issue timely warnings, but fortunately these iron-bound shores are not frequented by an amount of coasting craft at all to be compared with that navigating the comparatively calmer waters of the two Channels and the North Sea.

The results of the warnings *to our own coasts* have been printed as Parliamentary papers for several years back. The following is the abstract of these results for the year 1874 (Parl. Paper, No. 210, 1875, being the last which has been published).

Return of the Result of the Comparison between the Warnings issued and the Weather experienced in 1874.

Coasts	Total No. of orders to hoist and repetitions	Warnings justified by subsequent gales, force 8 and upwards	Warnings justified by subsequent strong winds, forces 6 and 7	Warnings not justified by subsequent weather	Warnings late, force 9 reached at two stations before issue	Warnings partially late, force 9 reached at one station before issue	Warnings late, owing to Sundays or telegraphic errors	Storms for which no warning was issued
Ireland, South	37	17	6	6	4	4	...	Apr. 13*s*, Sep. 22, Oct. 21
,, East	43	19	15	8	...	1	...	Jan. 18*s*, Sep. 22, Oct. 21
Scotland, East	35	15	9	10	...	1	...	Jan. 11*s*, Jan. 18*s*, Jan. 26*s*, Apr. 16, Aug. 6, Oct. 21, Dec. 8, Dec. 11
,, West (Clyde)	42	15	21	5	...	1	...	Jan. 18*s*, Mar. 27, Oct. 21
England, North-west	41	25	7	6	1	1	1	Oct. 21
,, West	37	19	12	5	...	...	1	Oct. 21
,, South	36	17	14	4	...	1	...	Oct. 21, Nov. 29
								Sep. 21*p*, at entrance of Channel
								Dec. 11*p*, in eastern portion of Channel
,, South-east	19	5	11	3	...	...	...	Oct. 21, Nov. 29
,, East	27	12	9	5	...	1	...	Jan. 18*s*, Oct. 21, Nov. 29
Totals	317	144	104	52	5	10	2	*s* indicates that the omission was owing to a Sunday, or to a telegraphic error
Per-centages	...	45·4	32·8	16·4	1·6	3·2	0·6	*p* indicates that the storm was only reported from one station on the line of coast

If these figures be compared with those for the year 1873, we see that the results in percentages are nearly identical.

	Warnings justified			Warnings not justified by subsequent weather
	By subsequent gales	By subsequent strong winds	Total success	
1873	45·2	34·0	**79·2**	16·8
1874	45·4	32·8	**78·2**	16·4

Accordingly it appears that nearly half the signals of approaching storms (force upwards of 8, Beaufort scale, 'a fresh gale') were fully justified, and the same is the case for four out of five signals of approaching strong winds (force upwards of 6, Beaufort scale, a 'strong breeze').

It is therefore evident that the system of storm warnings renders it possible to give useful intelligence to the coasts, but it need hardly be said that the system for these islands is rather of the 'from hand to mouth' type, and that our warnings would be far more useful if they could be issued sooner, so as to render it possible for captains to get intimation of a coming gale before leaving port. This, however, cannot be hoped for in the present state, the infancy, of weather telegraphy, and a few instances will suffice to show what are the principal causes of failure of warnings, in addition to those arising from the imperfections of our arrangements which have been already described.

I shall commence with a case in which even a

practised seaman's experience was at fault, so that we can hardly blame our telegraphic reporters, who are mostly landsmen, for their failure to notice from the appearance of the sky and sea, that a tremendous storm was imminent. Some of my readers may remember the gale of November 22, 1872, when the 'Royal Adelaide' was lost on Chesil Bank. On that day at noon the telegraphic reports showed an apparent improvement on the weather of the previous day, so that the signals then flying were lowered on the south coast. At night the storm came on, and of course the comments on the Office were not favourable. On this day one of our best sea-observers, the late Captain Thomas Donkin, of the 'Inverness,' (one of the three ships that rode out the Madras cyclone of May 2, 1872), was out in the gale, and was blown back by it, hove to, from the Lizard to the Casquets, off Alderney. I wrote to him as soon as I heard that he was safe at anchor at Portland, to ask him whether he, being at sea, had anticipated the storm from the look of the sky and sea, and his answer was: 'With respect to the weather on November 22, I may say that at noon I was standing in towards the land, between Falmouth and Plymouth, and a pilot cutter came alongside, and if I had had the least apprehension of such a gale as followed being near at hand, I should have taken a pilot and gone into Plymouth. The appearance of the weather at the time was fine, though the glass was falling, though not low at the time for SW. wind and unsettled weather.'

In this case, therefore, the signals, once hoisted, had been lowered too soon.

The next case, which I shall illustrate by charts, affords specimens of a double failure of warnings. In the first instance the signals had been hoisted on the north-west coasts on the morning of the 20th, and had been lowered again in the afternoon, while subsequently they were hoisted unnecessarily and had to be lowered. The gale is that of October 21, 1874, which has already been noticed p. 66.

Taking the first chart (fig. 46) for 8 a.m. October

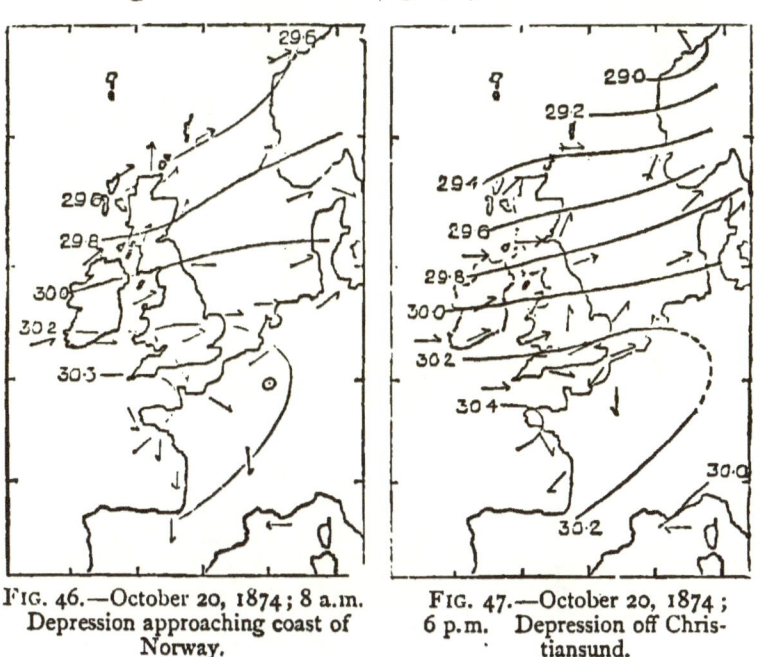

FIG. 46.—October 20, 1874; 8 a.m. Depression approaching coast of Norway.

FIG. 47.—October 20, 1874; 6 p.m. Depression off Christiansund.

20, we find pressure highest over the entrance of the Channel, and least in the far north, where SW. gales are reported. These gales, with a falling barometer in Scotland, caused some alarm, but this was allayed by finding conditions unchanged at 2 p.m. By 6 p.m., however, (fig. 47), the depression existing in the north had advanced to the coast of Norway, where the barometer had fallen 0·6 inch, but the signs of danger to ourselves were the bend southwards of the isobar of 29·4 round the Butt of Lewis, and the backing of the wind, though still light in force, to South-west at Stornoway and over the Moray Firth, and, though last not least, the steady rise of the barometer at the entrance of the Channel.

During the night the storm came on, and at 8 a.m. October 21, its centre lay half way between Peterhead and the Shetlands (fig. 48), while a terrific Westerly gale

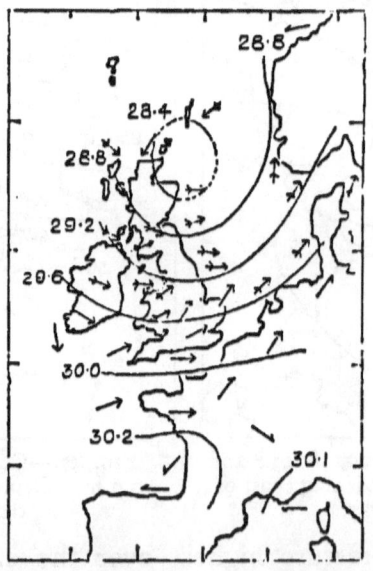

FIG. 48.—October 21, 1874; 8 a.m. New depression off north of Scotland.

was blowing over the whole of the United Kingdom, and the area of high pressure had been displaced, and lay over the extreme south-west of France.

The next chart (fig. 49), for 6 p.m. on the 21st, shows us that the disturbance had made rapid progress eastwards, and now lay over western Norway, while the winds in these islands had moderated in violence, and veered towards North-west.

If we now turn back to fig. 17, p. 77, we shall see the cause of the unnecessary warnings to which allusion has just been made.

At 8 a.m. on the 22nd, a brisk fall of the barometer took place in the west of Ireland, with a 'backing' wind, and warnings were at once issued, but as will be seen by fig. 17, this was only a false alarm, for the fresh disturbance was only subsidiary to the more serious one which had preceded it, and its only effect was to equalise pressure there, by reducing the gradients and moderating the force of the wind.

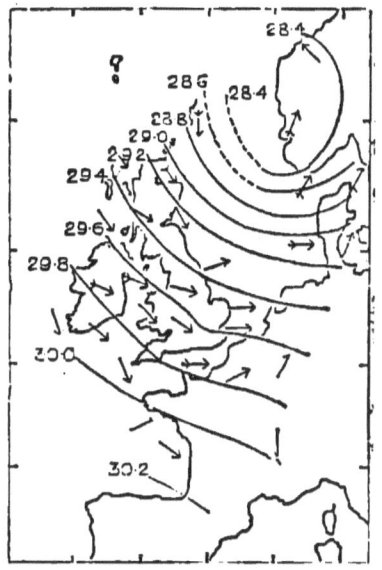

FIG. 49.—October 21, 1874; 6 p.m. Depression has advanced to the coast of Norway.

The storm which has been taken as a typical cyclone, that of November 29, 1874, whose course we have traced (figs. 19-22, pp. 83-85), is a good instance of a partial success and partial failure. On the 28th, warnings were issued to the west, north, and east of England, but not to the Channel coast or to the French stations. It will be seen from fig. 21 that while the warning was eminently successful on all the coasts to which it was issued, its omission was a striking failure at the southern stations.

It will be interesting to examine yet another case of failure of warnings, and one which has arisen from the unexpected direction of motion of the storm.

Fig. 50, for 8 a.m. January 21, 1875, shows us no

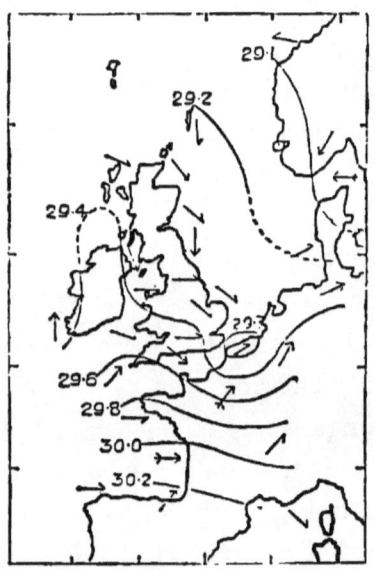

FIG. 50.—January 21, 1875; 8 a.m. Small depression over Belgium. New disturbance approaching Ireland.

less than three cyclonic disturbances over the area of our chart. One large one is disappearing over Norway. A very small subsidiary one is lying near Dover, while the third,—the one I am about to trace,—is just showing itself off the west coast of Ireland.

Warnings were issued to all our own coasts excepting Scotland and the north-west of England, from Fleetwood northwards. Fig. 51, for 2 p.m. on the same day, shows us the disturbance in question with its centre near Lundy Island, while pressure over the south of France has barely changed since the morning.

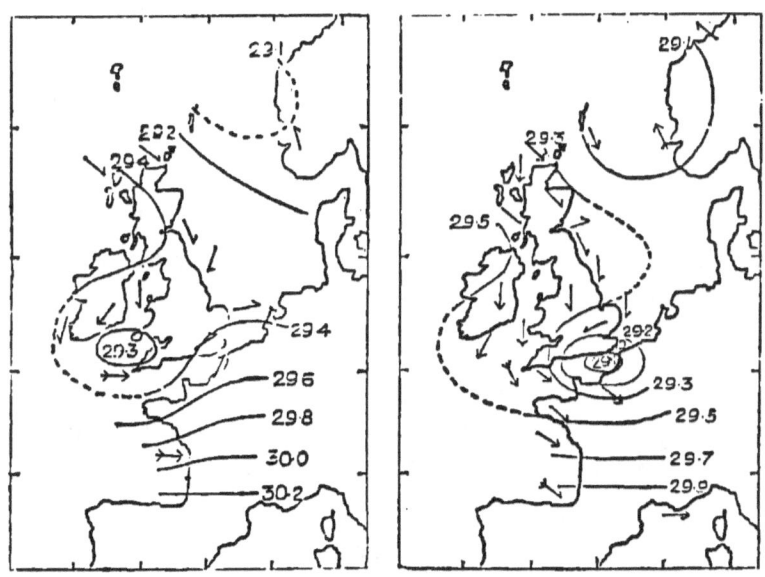

FIG. 51.—January 21, 1875; 2 p.m. New disturbance near Pembroke.

FIG. 52.—January 21, 1875; 6 p.m. New disturbance near Dover, in same position as that at 8 a.m.

At this time, however, the depression was becoming deeper, and was extending rapidly to the southward. Fig. 52, for 6 p.m. on the same day, shows us the centre of the storm lying over Dover, and a very heavy gale from North-west is blowing all along the west coast of France, while our shores had nothing but moderate breezes.

These warnings therefore were quite unnecessary for several of the English stations to which they were issued, but they ought to have been transmitted to France, where they would have been of good service.

Lastly, a storm like that of April 20-24, 1872 (p. 95), which doubles back on its own course, naturally causes much difficulty in the issue of warnings.

The instances cited are sufficient to show that there is great uncertainty in the practice of storm warnings, owing to the deficiency of our information as regards quality and quantity, and that even with all the information we can procure we are often to a great extent in the dark as to the character and motion, in direction and rate, of the storms which approach our coasts, until they have wrought more or less damage.

With all these drawbacks, however, I have shown at p. 137 that we are able to maintain a general success of nearly eighty per cent. for our warnings, a result on which we may fairly congratulate ourselves.

In conclusion, I can only say that the foregoing pages contain what is, necessarily, a very condensed and brief abstract of the present state of our knowledge of weather, in so far as this knowledge depends on the system of Weather Telegraphy, a branch of investigation which can hardly be said to have got out of the leading-strings of infancy as yet.

To do the subject justice, double the space, and at least treble the number of charts and illustrations, would have been required. If, however, the contents of these pages shall have awakened in any of my readers a real conception of what the weather work of the Meteorological Office is, and an interest in the careful study of weather, one of the most enticing of all enquiries—not to mention its practical importance to everyone—the trouble of putting together these notes will have been far more than repaid.

APPENDICES

APPENDIX A. p. 57.
SUNDAY, 8 A.M. WEATHER REPORT. NOVEMBER 29, 1874.

STATIONS	BAROMETER		THERMOMETERS (in shade)					WIND				Amount of cloud, 0 to 10	Weather, by Beaufort scale	Rainfall in past 24 hours	Sea disturbance, 0 to 9
	At 8 A.M. Reduced to 32° F. at the mean sea level	Change since yesterday	At 8 A.M.		Change in the dry bulb since yesterday	In past 24 hours		At 8 A.M.		Extreme					
			Dry bulb	Wet bulb		Max.	Min.	Direc.	Force (0 to 12)	Force (0 to 12)	Direc.				
Haparanda	29·82	−·36	11	11	−2	*	*	N.	2	2	N.	*	o	*	*
Hernösand	29·83	−·34	6	5	−12	*	*	Z.	0	2	N.	*	c o	*	*
Stockholm	29·82	−·28	30	30	−2	*	*	ENE.	2	2	E.	*	o s	*	*
Wisby	29·86	−·23	31	31	−2	*	*	SE.	4	0	E.	*	s c	*	*
Christiansund	29·71	−·31	27	25	−1	*	*	ESE.	2	6	Z.	*	b c	*	*
Skudesnaes	29·55	−·42	37	32	+1	*	*	E.	8	6	E.	*	c	*	3
Oxö (Christiansand)	29·60	−·42	37	34	+6	*	*	E.	6	4	NE.	*	s f	*	5
Skagen (the Scaw)	29·69	−·34	35	33	+1	*	*	ESE.	4	4	ESE.	7	o	*	4
Fanö	29·48	−·42	32	32	0	?	30	SE.	4	4	S.	7	o	·09	*
Cuxhaven	29·43 f	−·44	32	31	−2	43	38	SE.	5	5	SE.	10	o	·10	*
Sunburgh Head	29·32 f	−·27	41	41	−1	45	39	SE.	8	7	SE.	8	r p	·01	*
Stornoway (Hebrides)	29·03 f	−·37	42	39	−2	43	38	E.	6	6	SSE.	10	u	·10	8
Thurso	29·17 f	−·37	40	39	+1	44	38	ESE.	5	5	SSE.	8	u	·03	7
Wick	29·12 f	−·40	40	39	−1	47	36	SE.	7	7	S.	10	g u	···	3
Nairn	29·07 f	−·44	37	36	0	43	38	SE.	2	2	SE.	8	c o	·26	8
Aberdeen (9 a.m.)	29·02 f	−·56	41	39	+3	42	32	E.	7	7	SE.	8	r q	·07	*
Leith (9 a.m.)	28·87 f	−·69	40	40	+7	43	32	SE.	3	3	S.	10	o r	·40	8
Shields	28·81 f	−·82	41	41	+8	43	32	SE.	10	10	SE.	10	o r	·07	*
York	?	?	42	42	+10	43	34	E.	3	3	E.	10	r	·84	8

For explanation of the columns see p. 150.

Appendices. 149

Scarborough (10.30 a.m.)	28.59 f	−1.05	44	44	+10	44	34	SE.	10	9	SE.	10	o r	0.85	7
Nottingham	28.61 f	−1.03	45	45	+14	45	30	SSE.	2	7	ENE.	10	r	0.82	*
Ardrossan	28.78 f	−.73	38	40	+5	40	35	E.	6	6	E.	9	o h	0.04	3*
Greencastle (9 a.m.)	28.79 f	−.62	?49	?50	+6?	44	39	ESE.	2	7	E.	10	g d	0.20	5
Donaghadee	28.75 f	−.77	40	41	−2	43	41	NE.	6	7	ESE.	10	r d	1.12	4
Kingstown	28.67 f	−.78	42	41	−3	51	43	WNW.	6	6	SSE.	10	r	1.66	3
Holyhead	28.55 f	−.91	44	44	+3	46	40	ENE.	2	5	E.	10	r	0.87	:
Liverpool (Bidston Obs.)	28.57 f	−.97	40	41	+9	41	32	E.	3	7	N.	7	m f	0.74	7
Valencia	29.04 r	−.25	42	43	−5	50	42	N.	?	8	NW.	7	p q	0.32	5
Roche's Point	28.92 f	−.41	45	45	+2	51	42	NW.	6	8	WNW.	10	r c	0.45	7
Pembroke	28.68 f	−.75	43	46	+9	52	42	WNW.	10	10	S.	6	r q	1.19	3
Portishead	28.60 f	−.94	45	45	+1	45	34	S.	4	7	NW.	10	q u	0.76	7
Scilly	29.01 f	−.42	48	48	+5	54	44	WNW.	10	11	WSW.	6	o h q	0.45	4
Plymouth	28.81 f	−.65	47	48	+1	55	45	WNW.	6	8	WNW.	10	g q	0.81	6
Hurst Castle	28.65 f	−.90	49	49	+5	62	45	WNW.	7	7	S.	5	r q	1.17	5
Dover	28.80 f	−.77	50	50	+14	51	34	SSW.	6	5	SSE.	8	c u	0.07	*
London	28.63 f	−.99	49	49	+14	55	35	S.	2	5	SE.	10	g q o	0.71	*
Oxford	28.65	−.95	47	47	+12	51	35	S.	1	3	SSW.	10	r	0.93	*
Cambridge	28.67 f	−.96	46	47	+17	47	30	SSW.	2	2	SE.	10	g s o	0.67	5
Yarmouth	28.74 f	−.92	47	37	0	48	35	S.	5	5*	*	10*	g r	0.72	:
Helder	not received on Sundays		*	*	:	*	?	:	:	3	*	10	*	*	:
Cape Gris Nez	28.86 f	−.75	48	50	+23	?	?	S.	7	3	SE.	10	o r	0.36	5
Brest	?	:	50	52	:	56	47	W.	9	11	SW.	10	r f	0.47	8
L'Orient	29.15 f	−.34	47	51	+5	58	41	WNW.	9	9	WNW.	7	r	0.20	7
Rochefort	29.48 f	−.11	50	55	+10	?	43	WNW.	12	4	SE.	10	m o	0.43	5
Biarritz	29.75 r	+.13	54	56	+4	56	53	W.	7	5*	W.	10*	r o	0.16	7*
Corunna	30.00	:	*	*	:	*	*	SW.	6	*	*	*	o	*	*
Brussels	29.01	−.67	*	*	+12	*	*	SSE.	4	*	*	*	p o	0.20	*
Charleville	29.25	−.54	*	*	+11	*	*	S.	3	3	SE.	10	r c	*	*
Paris	29.15 f	−.49	46	43	+17	*	38	SSW.	7	9	*	10*	o	:	*
Lyons	?	?	49	51	+10	40	*	S.	6	4	E.	10	o	0.20	*
Toulon	29.79 f	+.01	48	48	+8	48	42	E.	1	5			r o	:	2

150   *Appendices.*

NOVEMBER 28, 2 P.M., REPORTS AND REMARKS.

Stations	Bar.	Dry	Wet	Wind	C.	Wea.	Sea
Skudesnaes .	29·87	37	33	?	6	*	o	4
Thurso . .	29·48	41	39	SSE.	5	8	o	3
Scarborough .	29·55	37	35	SSW.	2	10	o	3
Greencastle .	29·30	43	42	ESE.	5	10	g d	*
Holyhead . .	29·34	44	43	SSE.	2	10		1
Valencia . .	29·12	48	48	E.	2	8	r	5
Scilly . .	29·26	51	51	S.	6	10	d	6
London . .	29·53	39	38	SE.	2	10	o m g	*
Rochefort . .	29·60	50	48	SE.	4	10	m	3

The barometer continues to fall, excepting at Rochefort, most on our SW. coasts, and the wind at Valencia has backed to E. The sea is increasing at the mouth of the Channel, and rain falls along our western coasts. Warnings issued to our W. and N. coasts.

EXPLANATION OF COLUMNS.

BAROMETER.—The letters r (rising), f (falling), and s (steady), indicate the motion of the mercury in last 14 hours. EXTREME WIND is the strongest wind experienced in past 24 hours. WEATHER : see p. 18.

* *An Asterisk* is inserted in all places for which information is not usually received. The 'extreme' wind for the Swedish, Norwegian, and Danish stations is an observation taken at 8 p.m.

REMARKS

For Sunday, November 29, 1874, at 8 a.m.

During the past twenty-four hours a serious fall of the barometer has taken place over the United Kingdom, amounting to more than an inch at Nottingham ; and the centre of a very deep and well-defined depression has advanced to Wales and the western central parts of England. From Corunna to Pembroke there is a gradient of 0·13 inch per 50 miles for W. winds ; while from Skudesnaes to Holyhead there is one of 0·10 inch for SE. winds.

Temperature has risen very considerably at all but the most western stations. Over central England the change amounts to 14° in the twenty-four hours.

Strong SE. gales have set in on our NE. coast, while very strong W. to NW. gales are felt in the SW. and Bay of Biscay, and S. winds in the SE. of England. The sea runs very high, especially in the N. and NE.

Very heavy rain has fallen over the whole of the United Kingdom, accompanied by hail and snow in some places ; the fall continues at most stations. The weather is extremely unsettled.

NOTE.—The chart which faces this page is reduced from the Daily Weather Chart.

Weather Charts & Storm Warnings.

WEATHER CHARTS, Nov. 29. 1874, 8 A.M.
Reduced from the Daily Weather Report.

APPENDIX B.

Hourly readings of the self-recording instruments at Valencia Observatory, March 26, 27, 1874, p. 63.

MARCH 26.

Hour	Bar.	Temp. Dry	Temp. Wet	Tension of vapour	Wind Direction	Wind Velocity	Rain
1	...	...	...	...	...	...	...
2	...	...	...	...	...	...	...
3	...	...	...	...	...	...	...
4	...	...	...	...	...	...	...
5	...	...	...	...	...	...	...
6	...	...	...	...	...	...	...
7	...	...	...	...	...	...	...
8	...	...	...	...	...	...	...
9	30.004	50.9	49.2	.329	SSE.	39	...
10	29.990	50.8	49.4	.335	SSE.	37	...
11	29.978	51.0	49.8	.348	SSE.	42	...
Noon	29.924	50.8	50.0	.351	SSE.	40	0.005
1	29.874	50.7	50.0	.352	SSE.	46	0.005
2	29.874	50.6	49.7	.346	S. by E.	44	...
3	29.850	50.7	49.8	.347	SSE.	46	...
4	29.814	50.7	49.8	.347	S. by E.	51	0.050
5	29.806	50.6	49.9	.351	S. by E.	46	0.010
6	29.794	50.9	50.0	.349	S. by E.	45	...
7	29.778	51.4	50.7	.361	S.	50	...
8	29.772	51.6	50.9	.363	S. by E.	48	...
9	29.766	52.0	51.2	.366	S.	50	...
10	29.742	51.9	51.2	.367	S.	48	...

MARCH 27.

Hour	Bar.	Temp. Dry	Temp. Wet	Tension of vapour	Wind Direction	Wind Velocity	Rain
1	29.726	51.8	51.2	.369	S. by E.	46	...
2	29.714	52.0	51.3	.369	S. by E.	46	...
3	29.698	52.1	51.7	.379	S.	45	0.030
4	29.672	52.3	51.8	.379	S. by E.	41	0.020
5	29.650	52.2	51.8	.380	SSE.	39	0.030
6	29.634	52.0	51.6	.377	S. by E.	40	0.115
7	29.598	52.4	51.8	.377	S. by E.	44	0.065
8	29.564	52.7	51.9	.375	S.	48	...
9	29.546	52.6	50.7	.344	S. by W.	54	...
10	29.556	51.0	48.0	.332	SSW.	58	...
11	29.554	51.8	47.8	.299	SSW.	58	...
Noon	29.594	50.0	46.7	.285	SW. by S.	63	...
1	29.678	49.2	45.7	.280	WSW.	58	...
2	29.724	49.7	46.5	.268	W. by S.	44	0.010
3	29.768	51.0	46.2	.279	W.	34	...
4	29.816	49.0	45.9	.259	W.	33	...
5	29.874	48.9	44.0	.274	W. by N.	27	...
6	29.916	48.3	44.8	.236	WNW.	22	...
7	29.956	47.6	42.3	.257	WNW.	19	...
8	29.984	47.0	42.3	.215	WNW.	17	...
9	...	...	...	.220	WNW.	15	...
10	...	...	...	...	...	...	...
11	...	...	...	...	...	...	...
Midnight	...	...	...	...	...	...	...

APPENDIX C.

Hourly readings of the self-recording instruments at Aberdeen Observatory, October 20, 21, 1874, p. 67.

	OCTOBER 20.								OCTOBER 21.							
Hour	Bar.	Temp.		Tension of vapour	Wind		Rain	Hour	Bar.	Temp.		Tension of vapour	Wind		Rain
		Dry	Wet		Direction	Velocity				Dry	Wet		Direction	Velocity	
1	...	...	...	...	...	...	...	1	29.079	50.7	46.6	.271	SSW.	30	...
2	...	...	...	...	...	...	...	2	28.962	49.4	47.4	.303	S. by W.	28	...
3	...	...	...	...	...	...	...	3	28.827	50.1	48.0	.308	S.	28	...
4	...	...	...	...	...	...	...	4	28.657	50.1	48.2	.313	S. by W.	29	...
5	...	...	...	...	...	...	...	5	28.529	51.9	49.5	.324	SSW.	30	...
6	...	...	...	...	...	...	...	6	28.407	52.0	50.0	.335	SSW.	23	0.010
7	...	...	...	...	...	...	...	7	28.326	51.0	48.3	.306	SW. by S.	24	...
8	...	...	...	...	...	...	...	8	28.314	51.0	47.9	.297	SW. by W.	35	0.030
9	...	...	...	...	...	...	...	9	28.344	45.4	42.6	.241	W. by S.	46	0.020
10	...	...	...	...	...	...	...	10	28.419	42.5	41.4	.248	W.	55	0.010
11	...	...	...	...	...	...	...	11	28.644	40.8	40.8	.255	NW. by W.	57	0.180
Noon	...	...	...	...	SW. by S.	24	...	Noon	28.777	40.5	40.0	.241	WNW.	57	0.080
1	29.597	48.8	45.4	.265	SSW.	23	...	1	28.876	41.7	38.7	.204	WNW.	50	0.005
2	29.566	49.2	46.0	.274	SW.	22	...	2	28.988	41.0	38.0	.197	WNW.	41	0.005
3	29.551	50.1	46.5	.275	SW. by S.	14	...	3	28.995	40.8	37.4	.189	WNW.	42	0.005
4	29.546	50.5	46.9	.279	SSW.	14	...	4	29.061	41.0	37.6	.190	W.	32	...
5	29.536	49.0	46.4	.285	SSW.	11	...	5	29.095	39.3	36.1	.180	W. by N.	36	...
6	29.529	48.6	46.1	.282	SSW.	11	...	6	29.109	40.0	36.0	.172	W.	34	...
7	29.522	47.4	46.0	.293	SW. by S.	9	0.010	7	29.109	40.2	36.5	.179	W.	32	...
8	29.499	47.0	45.4	.284	SW. by S.	9	...	8	29.114	40.5	37.0	.185	W.	32	...
9	29.454	47.5	46.0	.292	SSW.	11	...	9	29.111	40.2	38.2	.209	W. by S.	30	0.015
10	29.408	48.0	46.5	.297	S. by W.	12	...	10	29.124	40.0	38.0	.207	W. by S.	27	...
11	29.341	48.9	47.3	.306	S.	14	...	11	29.128	42.1	39.3	.211	W. by N.	40	0.015
Midnight	29.254	50.5	48.5	.317	SSW.	24	...	Midnight	29.136	42.5	39.0	.202	W.	41	0.005
	29.172	51.3	48.1	.297	SSW.	35	...								

APPENDIX D.

Hourly readings of the self-recording instruments at Falmouth Observatory, Feb. 1, 2, 1873, p. 69.

FEBRUARY 1.

Hour	Bar.	Temp. Dry	Temp. Wet	Tension of vapour	Wind Direction	Wind Velocity	Rain
1	29·610	...	...	...	...	...	...
2	29·563	40·0	38·4	·212	...	...	...
3	29·520	39·6	38·4	·215	...	...	...
4	29·475	39·7	37·8	·219	...	...	...
5	29·430	39·0	37·2	·206	...	...	...
6	29·379	39·0	37·4	·203	...	...	...
7	29·331	39·0	37·4	·207	...	...	...
8	29·268	39·0	37·4	·207	...	...	...
9	29·223	38·0	37·0	·207	...	...	...
10	29·168	36·8	36·8	·208	...	...	0·010
11	29·129	37·0	36·0	·202	ESE.	42	·010
Noon	...	37·4	36·4	·200	ESE.	44	·055
1	29·048	39·4	38·3	·204	ESE.	46	·085
2	29·013	...	...	·220	SE. by E.	44	·060
3					ESE.	53	·075
4					ESE.	55	·070
5					ESE.	59	·090
6					ESE.	63	·140
7					ESE.	59	·150
8					E. by S.	64	·130
9					ESE.	62	·060
10					ESE.	65	·080
11					ESE.	63	
Midnight							

FEBRUARY 2.

Hour	Bar.	Temp. Dry	Temp. Wet	Tension of vapour	Wind Direction	Wind Velocity	Rain
1	28·948	42·2	41·2	·247	ESE.	51	0·070
2	28·923	45·2	44·3	·280	SE.	29	·040
3	28·896	45·7	44·3	·275	SSE.	23	·030
4	28·863	46·2	44·5	·274	S. by E.	26	·010
5	28·833	46·0	43·8	·261	S. by E.	22	·010
6	28·803	44·7	43·0	·257	SSE.	28	·020
7	28·773	44·7	42·5	·247	SSE.	29	...
8	28·720	44·7	42·3	·242	SE.	35	·035
9	28·708	44·9	42·4	·242	SE. by E.	37	...
10	28·667	44·5	42·3	·245	E. by S.	37	·035
11	28·637	43·9	42·0	·245	E. by N.	46	·010
Noon	28·631	43·8	42·4	·254	NE.	46	·020
1	28·661	44·2	42·8	·259	NE. by N.	46	·030
2	28·719	38·5	37·6	·215	N. by E.	40	·005
3	28·787	38·3	36·6	·198	N. by E.	36	·005
4	28·873	36·3	34·7	·185	N.	35	·005
5	28·953	35·0	33·3	·170	N. by W.	36	·005
6	29·039	34·9	32·7	·160	NNW.	35	...
7	29·108	34·7	32·4	·157	NNW.	35	...
8	29·161	34·6	32·3	·157	N. by W.	35	...
9	29·262	35·3	33·0	·162	N. by W.	33	...
10	29·262	35·8	33·9	·175	NNW.	33	...
11	29·303	36·0	33·7	·170	NNW.	29	...
Midnight	29·345	35·6	33·0	·160	NW. by N.	30	...

INDEX.

ABE

ABERDEEN Observatory, curves for October 21, 1874, 67; readings for same period, 152
America, possible value of telegraphic reports from, 126
Anticyclones defined, 32, 37; illustration of, 36; characteristics of, 53; typical, 54; motion of, 86
Atlantic Ocean, advance of storms over, 15, 110, 126
Aurora as sign of storm, 109
Azores, possible value of telegraphic reports from, 127

'BACKING' of wind explained, 26; reasons of, 72; why considered dangerous, 72, 129
Balaclava storm, experience of, 117
Barometer, corrections of, 2; — for capillarity, 2; — for capacity, 3; reduction of, to 32° F., 2; — to sea-level, 3; a single reading of, at any time, no indication of wind or weather, 23; scales, wording on, misleading, 44; rise and fall of, not infallible sign of weather, 49; instances of serious fall of, with calm weather, 50
Beaufort, Sir F., scale for wind, 7; notation of weather, 10
Belt, Thomas, F.G.S., theory of origin of cyclones, 111

CYC

'Brisk,' H.M.S., as signal-ship, 129
Buys Ballot, Prof., law of motion of wind, 22, 40

CHARTS, weather, explained, 33
Circulation of wind with cyclones, 35; with anticyclones, 38
Cirrus-clouds, value of, as signs of change of weather, 106
Clearness of atmosphere as sign of weather, 109
Clouds, deficiency of information respecting, in the Daily Weather Reports, 12; accompanying a cyclonic storm, 59
Coexisting depressions, 78
Currents, equatorial and polar, 20; relative positions of, as sign of disturbance, 131
Curves from self-recording instruments at times of storm, 63-69
Cyclones defined, 31, 53; illustrations of, 32, 53; typical, 57; passage of, on the northern side of a station, 61, 66; passage of, on the southern side of a station, 69; advance of, 61 *et seq.*; — probable rate of, unknown, 82; — affected by surface of ground, 88; — probable direction of, known to some extent, 88; —

affected by conditions of pressure, 89

DEPRESSIONS defined, 32; very extensive, instance of, 50; secondary, 74; co-existing, 78. *See* Cyclones
Draper, Prof. D., on storms crossing the Atlantic, 127

EVANS, LEWIS, first published statement of motion of storms, 80

FALMOUTH Observatory, curves for February 2, 1873, 69; readings for same period, 152
Faroe Islands, importance of information from, 128
Faye, M., theory of origin of storms, 112
FitzRoy, Admiral, institution of storm warnings, 118; storm-signal system, 119
Fogs with anticyclones, 57
Franklin, Benjamin, first recognition of motion of storms, 80

GERMANY, results of warnings from London to, 130
Gradients explained, 41, 49; instances of, 43, 48; for certain winds, 49; relation of, to winds, 51
Gyration of wind, law of, stated, 25; explained, 72

HAMBURG, warnings from London to, 130
Heart's Content, telegraphic reports from, value of, 126
Hoffmeyer's charts, 89

Humidity, importance of, as indication of weather, 6; changes of, during progress of storms, 65, 68, 70
Hurricanes, 40, 81

ISOBARS defined, 28; mode of drawing, 29

LAW, Buys Ballot's, 22, 40; of Gyration, 25, 71; of Storms, 40, 116
LeVerrier, M., first proposal of storm warnings in Europe, 117; ideas as to object of storm warnings, 117, 130
Ley, Rev. W. Clement, theory of motion of storms, 111
Loomis, Prof., on storms crossing the Atlantic, 126

MELDRUM, C., theory of origin of storms, 112; periodicity of rainfall, 113
Mistiness of atmosphere a sign of weather, 109
Mohn, Prof., theory of origin of storms, 111
Monsoons, 21
Motion of cyclones in different directions, effects of, 72; probable rate of, comparatively unknown, 82; in relation to mountains, 88; in relation to distribution of pressure, 89 *et seq.*; erratic, instance of, 95
Mountains, effects of, on motion of cyclones, 88; effects of, on winds, 125

NEWFOUNDLAND, telegraphic reports from, value of, 126

Index. 157

PRESSURE, relation of, to wind, 41; —— to motion of cyclones, 89

RAIN, not properly indicated by telegraphic weather reports, 7; with cyclones, 61; with easterly winds, 71
Rainfall, periodicity of, 113
Rate of motion of storms, 81
Redfield, discoverer of the Law of Storms, 116
Reid, Law of Storms, 116
Reports, telegraphic, defects of, 13
Results of Storm Warnings for the British Isles, 135
Reye, Prof., theory of origin of storms, 111

ST. KILDA, value of, as a possible reporting station, 128
Satellite depressions, instances of, 74
Sea disturbance, scale for, 11; value of, as indication of weather, 11
Sea-level, reduction of barometrical readings to, 3
Secondary depressions, instances of, 74; effects of, on their primaries, 75
Shower, with shift of wind to northwest, 60, 67
Signal-ships, proposal of, 15, 128
Signals, storm, explanation of, 119
Signs of storm, 131
Storms, definition of, 8; with high barometer, 23; vertical depth of, slight, 24; advance of, 81; rate of advance of, unknown, 82; probable direction of advance of, 87 *et seq.*; advance of, over the Atlantic, 15, 110, 126
Storm warnings, institution of, 118; difficulty of issuing correct, 132; results of, 135

Straight-line gales, explanation of, 24
Summary of principles recognised in weather study, 102
Sunday interruption to weather reports, 14
Sun-spots, relation of, to rainfall, 113

TEMPERATURE, correction of barometer for, 3; information respecting, given in Daily Weather Reports, 5; mean, definition of, 5; — mode of ascertaining, 5 instances of sudden falls of, 60; regular daily course of, disturbed by storms, 64, 68; differences of, as sign of storm, 134
Thermometer, attached, 2; wet and dry bulb, 6
Tornado in Sweden, 25
Trade winds, 21
Typhoons, 40, 81

VALENCIA Observatory, curves for March 27, 1874, 62; readings for same period, 151
Veering of wind explained, 26; reasons of, 72

WARNINGS, storm, first proposal of, in America, 116; first proposal of, in Europe, 117; first proposal of, in England, 118; results of, for Hamburg, 130; results of, for the United Kingdom, 135; instances of failure of, 137-144
Washington, Chief Signal Office, 14, 124
Weather, Beaufort notation for, 10; value of reports of general appearance of, 12; charts, explanation of, 32; with anticyclones, 53, 54; with cyclones, 53, 59;

WEA

charts, mode of using,. 106 ; local characteristics of, 107
Weather Reports in 1876, specimen of, 16 ; in 1874, specimen of, 148
Wind, Beaufort scale for, 7 ; cannot be measured instrumentally at telegraph stations, 8 ; measurement of, by velocity instead of pressure, 9 ; characteristics of main currents of, 19 ; currents,

WOR

polar and equatorial, 20 ; Trade, 21 ; Buys Ballot's law for, 22 ; motion of, in cyclones, 36 ; motion of, in anticyclones, 38 ; cause of, 41 ; relation of, to gradients, 49 ; reports of, from telegraphic stations sometimes deceptive, 125
Wording on barometer scales, misleading, 44

LONDON: PRINTED BY
SPOTTISWOODE AND CO., NEW-STREET SQUARE
AND PARLIAMENT STREET

THE
INTERNATIONAL SCIENTIFIC SERIES.

JOHN TYNDALL, LL.D., F.R.S.
The FORMS of WATER in CLOUDS and RIVERS, ICE and GLACIERS. With 26 Illustrations. Sixth Edition. Crown 8vo. cloth, price 5s. [Volume I.

WALTER BAGEHOT.
PHYSICS and POLITICS ; or, Thoughts on the Application of the Principles of Natural Selection and Inheritance to Political Society. Third Edition. Crown 8vo. cloth, price 4s. [Volume II.

EDWARD SMITH, M.D., LL.B., F.R.S. (the late).
FOODS. Profusely Illustrated. Fourth Edition. Crown 8vo. cloth, price 5s. [Volume III.

ALEXANDER BAIN, LL.D., Professor of Logic at the University of Aberdeen.
MIND and BODY : the Theories of their Relation. With Four Illustrations. Fifth Edition. Crown 8vo. cloth, price 4s. [Volume IV.

HERBERT SPENCER.
The STUDY of SOCIOLOGY. Fifth Edition. Crown 8vo. cloth, price 5s. [Volume V.

BALFOUR STEWART, M.A., LL.D., F.R.S.
The CONSERVATION of ENERGY ; being an Elementary Treatise on Energy and its Laws. Fourteen Engravings. Third Edition. Crown 8vo. cloth, price 5s. [Volume VI.

J. B. PETTIGREW, M.D., F.R.S.
ANIMAL LOCOMOTION : or, Walking, Swimming, and Flying. 130 Illustrations. Second Edition. Crown 8vo. cloth, price 5s. [Volume VII.

HENRY MAUDSLEY, M.D., Fellow of the Royal College of Physicians.
RESPONSIBILITY in MENTAL DISEASE. Second Edition. Crown 8vo. cloth, price 5s. [Volume VIII.

JOSIAH P. COOKE, Jun., Erving Professor of Chemistry and Mineralogy in Harvard College.
The NEW CHEMISTRY. 31 Illustrations. Third Edition. Crown 8vo. cloth, price 5s. [Volume IX.

SHELDON AMOS, M.A., Barrister-at-Law.
The SCIENCE of LAW. Second Edition. Crown 8vo. cloth, price 5s. [Volume X.

HENRY S. KING & CO., London.

The International Scientific Series.

E. J. MAREY, Professor at the College of France.
ANIMAL MECHANISM. 117 Illustrations. Second Edition. Crown 8vo. cloth, price 5s. [Volume XI.

OSCAR SCHMIDT, Professor in the University of Strasburg.
The DOCTRINE of DESCENT and DARWINISM. 26 Woodcuts. Third Edition. Crown 8vo. cloth, price 5s. [Volume XII.

JOHN WILLIAM DRAPER, M.D., LL.D.
HISTORY of the CONFLICT between RELIGION and SCIENCE. Seventh Edition. Crown 8vo. cloth, price 5s.
[Volume XIII.

M. C. COOKE, M.A., LL.D.
FUNGI : their Nature, Influences, and Uses. Edited by the Rev. M. J. BERKELEY, M.A., F.L.S. Profusely Illustrated. Second Edition. Crown 8vo. cloth, price 5s. [Volume XIV.

HERMANN VOGEL, Ph.D., Polytechnic Academy of Berlin.
The CHEMICAL EFFECTS of LIGHT and PHOTO-GRAPHY. 100 Illustrations. Third Edition, thoroughly revised. Crown 8vo. cloth, price 5s. [Volume XV.

WILLIAM DWIGHT WHITNEY, Professor of Sanskrit and Comparative Philology in Yale College.
The LIFE and GROWTH of LANGUAGE. Second Edition. Crown 8vo. cloth, price 5s. [Volume XVI.

W. STANLEY JEVONS, M.A., F.R.S.
MONEY and the MECHANISM of EXCHANGE. Third Edition. Crown 8vo. cloth, price 5s. [Volume XVII.

EUGENE LOMMEL, Professor of Physics in the University of Erlangen.
The NATURE of LIGHT, with a General Account of Physical Optics. 188 Illustrations. Second Edition. Crown 8vo. cloth, price 5s. [Vol. XVIII.

P. J. VAN BENEDEN, Professor at the University of Louvain.
ANIMAL PARASITES and MESSMATES. 83 Illustrations. Second Edition. Crown 8vo. cloth, price 5s. [Volume XIX.

P. SCHUTZENBERGER, Director of the Chemical Laboratory at Sorbonne.
On FERMENTATION. 28 Illustrations. Second Edition. Crown 8vo. cloth, price 5s. [Volume XX.

JULIUS BERNSTEIN, Professor of Physiology in the University of Halle.
The FIVE SENSES of MAN. 91 Woodcuts. Crown 8vo. cloth, price 5s. [Volume XXI.

HENRY S. KING & CO., London.

June, 1876.

AN ALPHABETICAL LIST

OF

HENRY S. KING & CO.'S
PUBLICATIONS.

65 Cornhill, and 1 Paternoster Square, London,
June, 1876.

A LIST OF

HENRY S. KING & CO.'S PUBLICATIONS.

ABBEY (Henry).
BALLADS OF GOOD DEEDS, AND OTHER VERSES. Fcap. 8vo. Cloth gilt, price 5s.

ADAMS (A. L.), M.A.
FIELD AND FOREST RAMBLES OF A NATURALIST IN NEW BRUNSWICK. With Notes and Observations on the Natural History of Eastern Canada. Illustrated. 8vo. Cloth, price 14s.

ADAMS (F. O.), H.B.M.'s Secretary of Embassy at Paris, formerly H.B.M.'s Chargé d'Affaires, and Secretary of Legation at Yedo.
THE HISTORY OF JAPAN. From the Earliest Period to the Present Time. New Edition, revised. In 2 vols. With Maps and Plans. Demy 8vo. Cloth, price 21s. each.

ADAMS (W. Davenport, Jun.)
LYRICS OF LOVE, from Shakespeare to Tennyson. Selected and arranged by. Fcap. 8vo. Cloth extra, gilt edges, price 3s. 6d.

ADON.
THROUGH STORM AND SUNSHINE. Illustrated by M. E. Edwards, A. T. H. Paterson, and the Author. Crown 8vo. Cloth, price 7s. 6d.

A. K. H. B.
A SCOTCH COMMUNION SUNDAY, to which are added Certain Discourses from a University City. By the Author of "The Recreations of a Country Parson." Second Edition. Crown 8vo. Cloth, price 5s.

ALLEN (Rev. R.), M.A.
 ABRAHAM: HIS LIFE, TIMES, AND TRAVELS, as told by a Contemporary 3800 years ago. With Map. Post 8vo. Cloth, price 10s. 6d.

AMOS (Professor Sheldon).
 THE SCIENCE OF LAW. Second Edition. Crown 8vo. Cloth, price 5s.
 Vol. X. of the International Scientific Series.

ANDERSON (Rev. Charles), M.A.
 NEW READINGS OF OLD PARABLES. Demy 8vo. Cloth, price 4s. 6d.
 CHURCH THOUGHT AND CHURCH WORK. Edited by. Containing articles by the Revs. J. M. Capes, Professor Cheetham, J. Ll. Davis, Harry Jones, Brooke, Lambert, A. J. Ross, the Editor, and others. Second Edition. Demy 8vo. Cloth, price 7s. 6d.
 WORDS AND WORKS IN A LONDON PARISH. Edited by. Second Edition. Demy 8vo. Cloth, price 6s.
 THE CURATE OF SHYRE. Second Edition. 8vo. Cloth, price 7s. 6d.

ANDERSON (Colonel R. P.)
 VICTORIES AND DEFEATS. An Attempt to explain the Causes which have led to them. An Officer's Manual. Demy 8vo. Cloth, price 14s.

ANDERSON (R. C.), C.E.
 TABLES FOR FACILITATING THE CALCULATION OF EVERY DETAIL IN CONNECTION WITH EARTHEN AND MASONRY DAMS. Royal 8vo. Cloth, price £2 2s.

ANSON (Lieut.-Col. The Hon. A.), V.C., M.P.
 THE ABOLITION OF PURCHASE AND THE ARMY REGULATION BILL OF 1871. Crown 8vo. Sewed, price 1s.
 ARMY RESERVES AND MILITIA REFORMS. Crown 8vo. Sewed, price 1s.
 THE STORY OF THE SUPERSESSIONS. Crown 8vo. Sewed, price 6d.

ARCHER (Thomas).
 ABOUT MY FATHER'S BUSINESS. Work amidst the Sick, the Sad, and the Sorrowing. Crown 8vo. Cloth, price 5s.

ARGYLE (Duke of).
 SPEECHES ON THE SECOND READING OF THE CHURCH PATRONAGE (SCOTLAND) BILL IN THE HOUSE OF LORDS, June 2, 1874; and Earl of Camperdown's Amendment, June 9, 1874, placing the Election of Ministers in the hands of Ratepayers. Crown 8vo. Sewed, price 1s.

ARMY OF THE NORTH GERMAN CONFEDERATION.
A Brief Description of its Organization, of the Different Branches of the Service and their *rôle* in War, of its Mode of Fighting, etc., etc. Translated from the Corrected Edition, by permission of the author, by Colonel Edward Newdegate. Demy 8vo. Cloth, price 5s.

ASHANTEE WAR (The).
A Popular Narrative. By the Special Correspondent of the Daily News. Crown 8vo. Cloth, price 6s.

ASHTON (John).
ROUGH NOTES OF A VISIT TO BELGIUM, SEDAN, AND PARIS, in September, 1870-71. Crown 8vo. Cloth, price 3s. 6d.

AUNT MARY'S BRAN PIE.
By the author of "St. Olave's," "When I was a Little Girl," etc. Illustrated. Cloth, price 3s. 6d.
SUNNYLAND STORIES. Illustrated. Fcap. 8vo. Cloth, price 3s. 6d.

AURORA.
A Volume of Verse. Fcap. 8vo. Cloth, price 5s.

AYRTON (J. C.)
A SCOTCH WOOING. 2 vols. Crown 8vo. Cloth.

BAGEHOT (Walter).
PHYSICS AND POLITICS ; or, Thoughts on the Application of the Principles of "Natural Selection" and "Inheritance" to Political Society. Third Edition. Crown 8vo. Cloth, price 4s.
Vol. II. of the International Scientific Series.
THE ENGLISH CONSTITUTION. A New Edition, Revised and Corrected, with an Introductory Dissertation on Recent Changes and Events. Crown 8vo. Cloth, price 7s. 6d.
LOMBARD STREET. A Description of the Money Market. Sixth Edition. Crown 8vo. Cloth, price 7s. 6d.

BAIN (Alexander), LL.D.
MIND AND BODY. The Theories of their Relation. Fifth Edition. Crown 8vo. Cloth, price 4s.
Vol. IV. of the International Scientific Series.

BANKS (Mrs. G. Linnæus).
GOD'S PROVIDENCE HOUSE. Crown 8vo. Cloth, price 3s. 6d.

BARING (T. C.), M.P., late Fellow of Brasenose College, Oxford.
PINDAR IN ENGLISH RHYME. Being an Attempt to render the Epinikian Odes with the principal remaining Fragments of Pindar into English Rhymed Verse. Small quarto. Cloth, price 7s.

BARLEE (Ellen).
> LOCKED OUT; A Tale of the Strike. With a Frontispiece. Cloth, price 1s. 6d.

BAYNES (Rev. Canon R. H.), Editor of "Lyra Anglicana," etc.
> HOME SONGS FOR QUIET HOURS. Second Edition. Fcap. 8vo. Cloth extra, price 3s. 6d.
> *₁* *This may also be had handsomely bound in Morocco with gilt edges.*

BECKER (Bernard H.)
> THE SCIENTIFIC SOCIETIES OF LONDON. Crown 8vo. Cloth, price 5s.

BENNETT (Dr. W. C.)
> SONGS FOR SAILORS. Dedicated by Special Request to H.R.H. the Duke of Edinburgh. With Steel Portrait and Illustrations. Crown 8vo. Cloth, price 3s. 6d.
> An Edition in Illustrated Paper Covers, price 1s.
> BABY MAY. HOME POEMS AND BALLADS. With Frontispiece. Crown 8vo. Cloth elegant, price 6s.
> BABY MAY AND HOME POEMS. Fcap. 8vo. Sewed in Coloured Wrapper, price 1s.
> NARRATIVE POEMS AND BALLADS. Fcap. 8vo. Sewed in Coloured Wrapper, price 1s.

BENNIE (Rev. Jas. Noble), M.A.
> THE ETERNAL LIFE. Sermons preached during the last twelve years. Crown 8vo. Cloth, price 6s.

BERNARD (Bayle).
> SAMUEL LOVER, THE LIFE AND UNPUBLISHED WORKS OF. In 2 vols. With a Steel Portrait. Post 8vo. Cloth, price 21s.

BERNSTEIN (Professor), of the University of Halle.
> THE FIVE SENSES OF MAN. With 91 Illustrations. Crown 8vo. Cloth, price 5s.
> Vol. XXI. of the International Scientific Series.

BETHAM-EDWARDS (Miss M.)
> KITTY. With a Frontispiece. Crown 8vo. Cloth, price 3s. 6d.
> MADEMOISELLE JOSEPHINE'S FRIDAYS, AND OTHER STORIES. Crown 8vo. Cloth, price 7s. 6d.

BISCOE (A. C.)
> THE EARLS OF MIDDLETON, Lords of Clermont and of Fettercairn, and the Middleton Family. Crown 8vo. Cloth, price 10s. 6d.

BLANC (Henry), M.D.
> CHOLERA: HOW TO AVOID AND TREAT IT. Popular and Practical Notes. Crown 8vo. Cloth, price 4s. 6d.

BLUME (Major William).
> THE OPERATIONS OF THE GERMAN ARMIES IN FRANCE, from Sedan to the end of the war of 1870-71. With Map. From the Journals of the Head-quarters Staff. Translated by the late E. M. Jones, Maj. 20th Foot, Prof. of Mil. Hist., Sandhurst. Demy 8vo. Cloth, price 9s.

BOGUSLAWSKI (Captain A. von).
> TACTICAL DEDUCTIONS FROM THE WAR OF 1870-71. Translated by Colonel Sir Lumley Graham, Bart., late 18th (Royal Irish) Regiment. Third Edition, Revised and Corrected. Demy 8vo. Cloth, price 7s.

BONWICK (James).
> THE TASMANIAN LILY. With Frontispiece. Crown 8vo. Cloth, price 5s.
> MIKE HOWE, THE BUSHRANGER OF VAN DIEMEN'S LAND. With Frontispiece. Crown 8vo. Cloth, price 5s.

BOSWELL (R. B.), M.A., Oxon.
> METRICAL TRANSLATIONS FROM THE GREEK AND LATIN POETS, and other Poems. Crown 8vo. Cloth, price 5s.

BOTHMER (Countess Von).
> CRUEL AS THE GRAVE. A Novel. 3 vols. Cloth.

BOWEN (H. C.), English Master Middle-Class City School, Cowper Street.
> STUDIES IN ENGLISH, for the use of Modern Schools. Small Crown 8vo. Cloth, price 1s. 6d.

BOWRING (L.), C.S.I., Lord Canning's Private Secretary, and for many years Chief Commissioner of Mysore and Coorg.
> EASTERN EXPERIENCES. Illustrated with Maps and Diagrams. Demy 8vo. Cloth, price 16s.

BRAVE MEN'S FOOTSTEPS.
> By the Editor of "Men who have Risen." A Book of Example and Anecdote for Young People. With Four Illustrations by C. Doyle. Third Edition. Crown 8vo. Cloth, price 3s. 6d.

BRIALMONT (Colonel A.)
> HASTY INTRENCHMENTS. Translated by Lieut. Charles A. Empsom, R.A. With nine Plates. Demy 8vo. Cloth, price 6s.

BRIEFS AND PAPERS.
Being Sketches of the Bar and the Press. By Two Idle Apprentices. Crown 8vo. Cloth, price 7s. 6d.

BROOKE (Rev. James M. S.), M. A.
HEART, BE STILL. A Sermon preached in Holy Trinity Church, Southall. Impl. 32mo. Sewed, price 6d.

BROOKE (Rev. Stopford A.), M.A., Chaplain in Ordinary to Her Majesty the Queen.
THE LATE REV. F. W. ROBERTSON, M.A., LIFE AND LETTERS OF. Edited by Stopford Brooke, M.A.
I. In 2 vols., uniform with the Sermons. Steel Portrait. Price 7s. 6d.
II. Library Edition. 8vo. Two Steel Portraits. Price 12s.
III. A Popular Edition, in 1 vol. 8vo. Price 6s.

THEOLOGY IN THE ENGLISH POETS.—COWPER, COLERIDGE, WORDSWORTH, and BURNS. Third Edition. Post 8vo. Cloth, price 9s.

CHRIST IN MODERN LIFE. Ninth Edition. Crown 8vo. Cloth, price 7s. 6d.

FREEDOM IN THE CHURCH OF ENGLAND. Six Sermons suggested by the Voysey Judgment. Second Edition. Crown 8vo. Cloth, price 3s. 6d.

SERMONS. First Series. Ninth Edition. Crown 8vo. Cloth, price 6s.

SERMONS. Second Series. Third Edition. Crown 8vo. Cloth, price 7s.

FREDERICK DENISON MAURICE: The Life and Work of. A Memorial Sermon. Crown 8vo. Sewed, price 1s.

BROOKE (W. G.), M.A., Barrister-at-Law.
THE PUBLIC WORSHIP REGULATION ACT. With a Classified Statement of its Provisions, Notes, and Index. Third Edition, revised and corrected. Crown 8vo. Cloth, price 3s. 6d.

SIX PRIVY COUNCIL JUDGMENTS—1850-1872. Annotated by. Third Edition. Crown 8vo. Cloth, price 9s.

BROWN (Rev. J. Baldwin), B.A.
THE HIGHER LIFE. Its Reality, Experience, and Destiny. Fourth Edition. Crown 8vo. Cloth, price 7s. 6d.

THE DOCTRINE OF ANNIHILATION IN THE LIGHT OF THE GOSPEL OF LOVE. Five Discourses. Second Edition. Crown 8vo. Cloth, price 2s. 6d.

BROWN (John Croumbie), LL.D., etc.
  REBOISEMENT IN FRANCE; or, Records of the Replanting of the Alps, the Cevennes, and the Pyrenees with Trees, Herbage, and Bush. Demy 8vo. Cloth, price 12s. 6d.
  THE HYDROLOGY OF SOUTHERN AFRICA. Demy 8vo. Cloth, price 10s. 6d.

BROWNE (Rev. Marmaduke E.)
  UNTIL THE DAY DAWN. Four Advent Lectures delivered in the Episcopal Chapel, Milverton, Warwickshire, on the Sunday evenings during Advent, 1870. Crown 8vo. Cloth, price 2s. 6d.

BRYANT (William Cullen).
  POEMS. Red-line Edition. With 24 Illustrations and Portrait of the Author. Post 8vo. Cloth extra, price 7s. 6d.
  A Cheaper Edition, with Frontispiece. Post 8vo. Cloth, price 3s. 6d.

BUCHANAN (Robert).
  POETICAL WORKS. Collected Edition, in 3 Vols., with Portrait. Price 6s. each.
  CONTENTS OF THE VOLUMES.
  I. "Ballads and Romances." II. "Ballads and Poems of Life." III. "Cruiskeen Sonnets;" and "Book of Orm."
  MASTER-SPIRITS. Post 8vo. Cloth, price 10s. 6d.

BULKELEY (Rev. Henry J.)
  WALLED IN, and other Poems. Crown 8vo. Cloth, price 5s.

BUNNÈTT (F. E.)
  LEONORA CHRISTINA, MEMOIRS OF, Daughter of Christian IV. of Denmark; Written during her Imprisonment in the Blue Tower of the Royal Palace at Copenhagen, 1663-1685. Translated by F. E. Bunnètt. With an Autotype Portrait of the Princess. A New and Cheaper Edition. Medium 8vo. Cloth, price 5s.
  LINKED AT LAST. 1 vol. Crown 8vo. Cloth.
  UNDER A CLOUD; OR, JOHANNES OLAF. By E. D. Wille. Translated by F. E. Bunnètt. 3 vols. Crown 8vo. Cloth.

BURTON (Mrs. Richard).
  THE INNER LIFE OF SYRIA, PALESTINE, AND THE HOLY LAND. 2 vols. Second Edition. Demy 8vo. Cloth, price 24s.

BUTLER (Josephine E.)
  JOHN GREY (of Dilston): MEMOIRS. By his Daughter. New and Cheaper Edition. Crown 8vo. Cloth, price 3s. 6d.

CADELL (Mrs. H. M.)
    IDA CRAVEN: A Novel. 2 vols. Crown 8vo. Cloth.

CALDERON.
    CALDERON'S DRAMAS: The Wonder-Working Magician—Life is a Dream—The Purgatory of St. Patrick. Translated by Denis Florence MacCarthy. Post 8vo. Cloth, price 10s.

CARLISLE (A. D.), B.A., Trin. Coll., Camb.
    ROUND THE WORLD IN 1870. A Volume of Travels, with Maps. New and Cheaper Edition. Demy 8vo. Cloth, price 6s.

CARNE (Miss E. T.)
    THE REALM OF TRUTH. Crown 8vo. Cloth, price 5s. 6d.

CARPENTER (E.)
    NARCISSUS AND OTHER POEMS. Fcap. 8vo. Cloth, price 5s.

CARPENTER (W. B.), LL.D., M.D., F.R.S., etc.
    THE PRINCIPLES OF MENTAL PHYSIOLOGY. With their Applications to the Training and Discipline of the Mind, and the Study of its Morbid Conditions. Illustrated. 8vo. Fourth Edition. Cloth, price 12s.

CARR (Lisle).
    JUDITH GWYNNE. 3 vols. Second Edition. Crown 8vo. Cloth.

CHRISTOPHERSON (The late Rev. Henry), M.A., Assistant Minister at Trinity Church, Brighton.
    SERMONS. With an Introduction by John Rae, LL.D., F.S.A. First Series. Crown 8vo. Cloth, price 7s. 6d.
    SERMONS. With an introduction by John Rae, LL.D., F.S.A. Second Series. Crown 8vo. Cloth, price 6s.

CLAYTON (Cecil).
    EFFIE'S GAME; HOW SHE LOST AND HOW SHE WON. A Novel. 2 vols. Cloth.

CLERK (Mrs. Godfrey), Author of "The Antipodes and Round the World."
    'ILAM EN NAS. Historical Tales and Anecdotes of the Times of the Early Khalifahs. Translated from the Arabic Originals. Illustrated with Historical and Explanatory Notes. Crown 8vo. Cloth, price 7s.

CLERY (C.), Captain 32nd Light Infantry, Deputy Assistant Adjutant-General, late Professor of Tactics Royal Military College, Sandhurst.
    MINOR TACTICS. With 26 Maps and Plans. Second Edition. Demy 8vo. Cloth, price 16s.

CLODD (Edward), F.R.A.S.
> **THE CHILDHOOD OF THE WORLD:** a Simple Account of Man in Early Times. Third Edition. Crown 8vo. Cloth, price 3s.
>
> A Special Edition for Schools. Price 1s.
>
> **THE CHILDHOOD OF RELIGIONS.** Including a Simple Account of the Birth and Growth of Myths and Legends. Crown 8vo. Cloth, price 5s.

COLERIDGE (Sara).
> **PRETTY LESSONS IN VERSE FOR GOOD CHILDREN**, with some Lessons in Latin, in Easy Rhyme. A New Edition. Illustrated. Cloth, price 3s. 6d.
>
> **PHANTASMION.** A Fairy Romance. With an Introductory Preface by the Right Hon. Lord Coleridge, of Ottery St. Mary. A New Edition. Illustrated. Cloth, price 7s. 6d.
>
> **MEMOIR AND LETTERS OF SARA COLERIDGE.** Edited by her Daughter. Third Edition, Revised and Corrected. With Index. 2 vols. With Two Portraits. Crown 8vo. Cloth, price 24s.
>
> Cheap Edition. With one Portrait. Cloth, price 7s. 6d.

COLLINS (Mortimer).
> **THE PRINCESS CLARICE.** A Story of 1871. 2 vols. Cloth.
>
> **SQUIRE SILCHESTER'S WHIM.** 3 vols. Cloth.
>
> **MIRANDA.** A Midsummer Madness. 3 vols. Cloth.
>
> **THE INN OF STRANGE MEETINGS, AND OTHER POEMS.** Crown 8vo. Cloth, price 5s.
>
> **THE SECRET OF LONG LIFE.** Dedicated by special permission to Lord St. Leonard's. Fourth Edition. Large crown 8vo. Price 5s.

COLLINS (Rev. Richard), M.A.
> **MISSIONARY ENTERPRISE IN THE EAST.** With special reference to the Syrian Christians of Malabar, and the results of modern Missions. With Four Illustrations. Crown 8vo. Cloth, price 6s.

CONGREVE (Richard), M.A., M.R.C.P.L.
> **HUMAN CATHOLICISM.** Two Sermons delivered at the Positivist School on the Festival of Humanity, 87 and 88, January 1, 1875 and 1876. Demy 8vo. Sewed, price 1s.

CONWAY (Moncure D.)
> **REPUBLICAN SUPERSTITIONS.** Illustrated by the Political History of the United States. Including a Correspondence with M. Louis Blanc. Crown 8vo. Cloth, price 5s.

CONYERS (Ansley).
> **CHESTERLEIGH.** 3 vols. Crown 8vo. Cloth.

COOKE (M. C.), M.A., LL.D.
> FUNGI; their Nature, Influences, Uses, etc. Edited by the Rev. M. J. Berkeley, M.A., F.L.S. With Illustrations. Second Edition. Crown 8vo. Cloth, price 5s.
> Vol. XIV. of the International Scientific Series.

COOKE (Professor Josiah P.), of the Harvard University.
> THE NEW CHEMISTRY. With Thirty-one Illustrations. Third Edition. Crown 8vo. Cloth, price 5s.
> Vol. IX. of the International Scientific Series.
> SCIENTIFIC CULTURE. Crown 8vo. Cloth, price 1s.

COOPER (T. T.)
> THE MISHMEE HILLS: an Account of a Journey made in an Attempt to Penetrate Thibet from Assam, to open New Routes for Commerce. Second Edition. With Four Illustrations and Map. Demy 8vo. Crown 8vo. Cloth, price 10s. 6d.

CORNHILL LIBRARY OF FICTION (The.) Crown 8vo. Cloth, price 3s. 6d. per Volume.
> HALF-A-DOZEN DAUGHTERS. By J. Masterman.
> THE HOUSE OF RABY. By Mrs. G. Hooper.
> A FIGHT FOR LIFE. By Moy Thomas.
> ROBIN GRAY. By Charles Gibbon.
> KITTY. By Miss M. Betham-Edwards.
> HIRELL. By John Saunders.
> ONE OF TWO; OR, THE LEFT-HANDED BRIDE. By J. Hain Friswell.
> READY-MONEY MORTIBOY. A Matter-of-Fact Story.
> GOD'S PROVIDENCE HOUSE. By Mrs. G. L. Banks.
> FOR LACK OF GOLD. By Charles Gibbon.
> ABEL DRAKE'S WIFE. By John Saunders.

CORY (Lieutenant-Colonel Arthur).
> THE EASTERN MENACE; OR, SHADOWS OF COMING EVENTS. Crown 8vo. Cloth, price 5s.

COSMOS.
> A Poem. Fcap. 8vo. Cloth, price 3s. 6d.
> SUBJECTS.—Nature in the Past and in the Present—Man in the Past and in the Present—The Future.

COTTON (Robert Turner).
> MR. CARINGTON. A Tale of Love and Conspiracy. 3 vols. Crown 8vo. Cloth.

CUMMINS (Henry Irwin), M.A.
> PAROCHIAL CHARITIES OF THE CITY OF LONDON. Sewed, price 1s.

CURWEN (Henry).
SORROW AND SONG: Studies of Literary Struggle. Henry Mürger—Novalis—Alexander Petöfi—Honoré de Balzac—Edgar Allan Poe—André Chénier. 2 vols. Crown 8vo. Cloth, price 15s.

DAVIDSON (Samuel), D.D., LL.D.
THE NEW TESTAMENT, TRANSLATED FROM THE LATEST GREEK TEXT OF TISCHENDORF. A new and thoroughly revised Edition. Post 8vo. Cloth, price 10s. 6d.

DAVIES (G. Christopher).
MOUNTAIN, MEADOW, AND MERE: a Series of Outdoor Sketches of Sport, Scenery, Adventures, and Natural History. With Sixteen Illustrations by Bosworth W. Harcourt. Crown 8vo. Cloth, price 6s.
RAMBLES AND ADVENTURES OF OUR SCHOOL FIELD CLUB. With 4 Illustrations. Crown 8vo. Cloth, price 5s.

DAVIES (Rev. J. Llewelyn), M.A.
THEOLOGY AND MORALITY. Essays on Questions of Belief and Practice. Crown 8vo. Cloth, price 7s. 6d.

D'ANVERS (N. R.)
THE SUEZ CANAL: Letters and Documents descriptive of its Rise and Progress. By Ferdinand de Lesseps. Translated by N. D'Anvers. Demy 8vo. Cloth, price 10s. 6d.
LITTLE MINNIE'S TROUBLES. An Every-day Chronicle. Illustrated by W. H. Hughes. Fcap. Cloth, price 3s. 6d.

DANCE (Rev. Charles Daniel).
RECOLLECTIONS OF FOUR YEARS IN VENEZUELA. With Three Illustrations and a Map. Crown 8vo. Cloth, price 7s. 6d.

DE KERKADEC (Vicomtesse Solange).
A CHEQUERED LIFE, being Memoirs of the Vicomtesse de Leoville Meilhan. Edited by. Crown 8vo. Cloth, price 7s. 6d.

DE L'HOSTE (Colonel E. P.).
THE DESERT PASTOR, JEAN JAROUSSEAU. Translated from the French of Eugène Pelletan. With a Frontispiece. New Edition. Fcap. 8vo. Price 3s. 6d.

DE REDCLIFFE (Viscount Stratford), P.C., K.G., G.C.B.
WHY AM I A CHRISTIAN? Fifth Edition. Crown 8vo. Cloth, price 3s.

DE TOCQUEVILLE (Alexis).
CORRESPONDENCE AND CONVERSATIONS OF, WITH NASSAU WILLIAM SENIOR. 2 vols. Post 8vo. Cloth, price 21s.

DE VERE (Aubrey).
ST. THOMAS OF CANTERBURY. A Dramatic Poem. Large fcap. 8vo. Cloth, price 5s.
ALEXANDER THE GREAT. A Dramatic Poem. Small crown 8vo. Cloth, price 5s.
THE INFANT BRIDAL, AND OTHER POEMS. A New and Enlarged Edition. Fcap. 8vo. Cloth, price 7s. 6d.
THE LEGENDS OF ST. PATRICK, AND OTHER POEMS. Small crown 8vo. Cloth, price 5s.

DE WILLE (E.)
UNDER A CLOUD; OR, JOHANNES OLAF. A Novel. Translated by F. E. Bunnètt. 3 vols. Crown 8vo. Cloth.

DENNIS (John).
ENGLISH SONNETS. Collected and Arranged. Elegantly bound. Fcap. 8vo. Cloth, price 3s. 6d.

DOBSON (Austin).
VIGNETTES IN RHYME AND VERS DE SOCIÉTÉ. Second Edition. Fcap. 8vo. Cloth, price 5s.

DONNÉ (Alphonse), M.D.
CHANGE OF AIR AND SCENE. A Physician's Hints about Doctors, Patients, Hygiene, and Society; with Notes of Excursions for Health in the Pyrenees, and amongst the Wateringplaces of France (Inland and Seaward), Switzerland, Corsica, and the Mediterranean. A New Edition. Large post 8vo. Cloth, price 9s.

DOWDEN (Edward), LL.D.
SHAKSPERE: a Critical Study of his Mind and Art. Second Edition. Post 8vo. Cloth, price 12s.

DOWNTON (Rev. Henry), M.A.
HYMNS AND VERSES. Original and Translated. Small crown 8vo. Cloth, price 3s. 6d.

DRAPER (John William), M.D., LL.D. Professor in the University of New York; Author of "A Treatise on Human Physiology."
HISTORY OF THE CONFLICT BETWEEN RELIGION AND SCIENCE. Seventh Edition. Crown 8vo. Cloth, price 5s.
Vol. XIII. of the International Scientific Series.

DREW (Rev. G. S.), M.A., Vicar of Trinity, Lambeth.
SCRIPTURE LANDS IN CONNECTION WITH THEIR HISTORY. Second Edition. 8vo. Cloth, price 10s. 6d.
NAZARETH: ITS LIFE AND LESSONS. Third Edition. Crown 8vo. Cloth, price 5s.
THE DIVINE KINGDOM ON EARTH AS IT IS IN HEAVEN. 8vo. Cloth, price 10s. 6d.
THE SON OF MAN: His Life and Ministry. Crown 8vo. Cloth, price 7s. 6d.

DREWRY (G. Overend), M.D.
THE COMMON-SENSE MANAGEMENT OF THE STOMACH. Third Edition. Fcap. 8vo. Cloth, price 2s. 6d.
CUP AND PLATTER; OR, NOTES ON FOOD AND ITS EFFECTS. By G. O. Drewry, M.D., Author of "The Common-Sense Management of the Stomach," and H. C. Bartlett, Ph.D., F.C.S. Fcap. 8vo. Cloth, price 2s. 6d.

DURAND (Lady).
IMITATIONS FROM THE GERMAN OF SPITTA AND TERSTEGEN. Fcap. 8vo. Cloth, price 4s.

DU VERNOIS (Colonel von Verdy).
STUDIES IN LEADING TROOPS. An authorized and accurate Translation by Lieutenant H. J. T. Hildyard, 71st Foot. Parts I. and II. Demy 8vo. Cloth, price 7s.

E. A. V.
JOSEPH MAZZINI: A Memoir. With Two Essays by Mazzini—"Thoughts on Democracy," and "The Duties of Man." Dedicated to the Working Classes by P. H. Taylor, M.P. With Two Portraits. Crown 8vo. Cloth, price 3s. 6d.

EDEN (Frederic).
THE NILE WITHOUT A DRAGOMAN. Second Edition. Crown 8vo. Cloth, price 7s. 6d.

EDWARDS (Rev. Basil).
MINOR CHORDS; OR, SONGS FOR THE SUFFERING: a Volume of Verse. Fcap. 8vo. Cloth, price 3s. 6d.; paper, price 2s. 6d.

EILOART (Mrs.)
LADY MORETOUN'S DAUGHTER. 3 vols. Crown 8vo.

ENGLISH CLERGYMAN.
AN ESSAY ON THE RULE OF FAITH AND CREED OF ATHANASIUS. Shall the Rubric preceding the Creed be removed from the Prayer-book? Sewed. 8vo. Price 1s.

EROS AGONISTES.
    Poems. By E. B. D. Fcap. 8vo. Cloth, price 3s. 6d.

ESSAYS ON THE ENDOWMENT OF RESEARCH.
    By Various Writers.
    LIST OF CONTRIBUTORS.

MARK PATTISON, B.D.
JAMES S. COTTON, B.A.
CHARLES E. APPLETON, D.C.L.
ARCHIBALD H. SAYCE, M.A.

HENRY CLIFTON SORBY, F.R.S.
THOMAS K. CHEYNE, M.A.
W. T. THISTELTON DYER, M.A.
HENRY NETTLESHIP, M.A.

    Square crown octavo. Cloth, price 10s. 6d.

EVANS (Mark).
    THE STORY OF OUR FATHER'S LOVE, told to Children: being a New and Enlarged Edition of THEOLOGY FOR CHILDREN. Fcap. 8vo. Cloth, price 3s. 6d.
    A BOOK OF COMMON PRAYER AND WORSHIP FOR HOUSEHOLD USE, compiled exclusively from the Holy Scriptures. Fcap. 8vo. Cloth, price 2s. 6d.

EYRE (Maj.-Gen. Sir Vincent), C.B., K.C.S.I., etc.
    LAYS OF A KNIGHT-ERRANT IN MANY LANDS. Square crown 8vo. With Six Illustrations. Cloth, price 7s. 6d.

FAITHFULL (Mrs. Francis G.)
    LOVE ME, OR LOVE ME NOT. 3 vols. Crown 8vo. Cloth.

FARQUHARSON (Martha).
    I. **ELSIE DINSMORE.** Crown 8vo. Cloth, price 3s. 6d.
    II. **ELSIE'S GIRLHOOD.** Crown 8vo. Cloth, price 3s. 6d.
    III. **ELSIE'S HOLIDAYS AT ROSELANDS.** Crown 8vo. Cloth, price 3s. 6d.

FAVRE (Mons. Jules).
    THE GOVERNMENT OF THE NATIONAL DEFENCE. From the 30th June to the 31st October, 1870. The Plain Statement of a Member. Demy 8vo. Cloth, price 10s. 6d.

FISHER (Alice).
    HIS QUEEN. 3 vols. Crown 8vo. Cloth.

FOOTMAN (Rev. Henry).
    FROM HOME AND BACK; or, some Aspects of Sin as Seen in the Light of the Parable of the Prodigal. Crown 8vo. Cloth, price 5s.

FORBES (Archibald).
    SOLDIERING AND SCRIBBLING. A Series of Sketches. Crown 8vo. Cloth, price 7s. 6d.

FOTHERGILL (JESSIE).
    HEALEY. A Romance. 3 vols. Crown 8vo. Cloth.

FOWLE (Rev. T. W.), M.A.
THE RECONCILIATION OF RELIGION AND SCIENCE. Being Essays on Immortality, Inspiration, Miracles, and the Being of Christ. Demy 8vo. Cloth, price 10s. 6d.

FOX-BOURNE (H. R.)
THE LIFE OF JOHN LOCKE, 1632—1704. 2 vols. Demy 8vo. Cloth, price 28s.

FRASER (Donald), Accountant to the British-Indian Steam Navigation Company, Limited.
EXCHANGE TABLES OF STERLING AND INDIAN RUPEE CURRENCY, upon a new and extended system, embracing Values from One Farthing to One Hundred Thousand Pounds, and at Rates progressing, in Sixteenths of a Penny, from 1s. 9d. to 2s. 3d. per Rupee. Royal 8vo. Cloth, price 10s. 6d.

FRERE (Sir H. Bartle E.), G.C.B., G.C.S.I., etc.
THE THREATENED FAMINE IN BENGAL: How it may be Met, and the Recurrence of Famines in India Prevented. Being No. 1 of "Occasional Notes on Indian Affairs." With 3 Maps. Crown 8vo. Price 5s.

FRISWELL (J. Hain).
THE BETTER SELF. Essays for Home Life. Crown 8vo. Price 6s.
ONE OF TWO; OR, THE LEFT-HANDED BRIDE. With a Frontispiece. Crown 8vo. Price 3s. 6d.

GARDNER (John), M.D.
LONGEVITY; THE MEANS OF PROLONGING LIFE AFTER MIDDLE AGE. Third Edition, revised and enlarged. Small crown 8vo. Cloth, price 4s.

GARDNER (Herbert).
SUNFLOWERS. A Book of Verses. Fcap. 8vo. Cloth, price 5s.

GARRETT (Edward).
BY STILL WATERS. A Story for Quiet Hours. With Seven Illustrations. Crown 8vo. Cloth, price 6s.

GIBBON (Charles).
FOR LACK OF GOLD. With a Frontispiece. Crown 8vo. Cloth, price 3s. 6d.
ROBIN GRAY. With a Frontispiece. Crown 8vo. Cloth, price 3s. 6d.

GILBERT (Mrs.)
MRS. GILBERT, FORMERLY ANN TAYLOR, AUTOBIOGRAPHY AND OTHER MEMORIALS OF. Edited by Josiah Gilbert. New and revised Edition. In 2 vols. With 2 Steel Portraits and several Wood Engravings. Post 8vo. Cloth, price 24s.

B   b

GILL (Rev. W. W.), B.A., of the London Missionary Society.
> **MYTHS AND SONGS FROM THE SOUTH PACIFIC.** With a Preface by F. Max Müller, M.A., Professor of Comparative Philology at Oxford. Post 8vo. Cloth, price 9s.

GODKIN (James).
> **THE RELIGIOUS HISTORY OF IRELAND:** Primitive, Papal, and Protestant. Including the Evangelical Missions, Catholic Agitations, and Church Progress of the last half Century. 8vo. Cloth, price 12s.

GODWIN (William).
> **WILLIAM GODWIN: HIS FRIENDS AND CONTEMPORARIES.** With Portraits and Facsimiles of the handwriting of Godwin and his Wife. By C. Kegan Paul. 2 vols. Demy 8vo. Cloth, price 28s.
>
> **THE GENIUS OF CHRISTIANITY UNVEILED.** Being Essays never before published. Edited, with a Preface, by C. Kegan Paul. Crown 8vo. Cloth, price 7s. 6d.

GOETZE (Capt. A. von), Captain of the Prussian Corps of Engineers attached to the Engineer Committee, and Instructor at the Military Academy.
> **OPERATIONS OF THE GERMAN ENGINEERS DURING THE WAR OF 1870-1871.** Published by Authority, and in accordance with Official Documents. Translated from the German by Colonel G. Graham, V.C., C.B., R.E. With 6 large Maps. Demy 8vo. Cloth, price 21s.

GOODENOUGH (Commodore J. G.), Royal Navy, C.B., C.M.G.
> **JOURNALS OF,** during his Last Command as Senior Officer on the Australian Station, 1873-1875. Edited, with a Memoir, by his Widow. With Maps, Woodcuts, and Steel Engraved Portrait. Crown 8vo. Cloth, price 14s.

GOODMAN (Walter).
> **CUBA, THE PEARL OF THE ANTILLES.** Crown 8vo. Cloth, price 7s. 6d.

GOSSE (Edmund W.)
> **ON VIOL AND FLUTE.** With Title-page specially designed by William B. Scott. Crown 8vo. Cloth, price 5s.

GOULD (Rev. S. Baring).
> **THE VICAR OF MORWENSTOW:** a Memoir of the Rev. R. S. Hawker. With Portrait. Second Edition, revised. Post 8vo. Cloth, price 10s. 6d.

GRANVILLE (A. B.), M.D., F.R.S., etc.
> AUTOBIOGRAPHY OF A. B. GRANVILLE, F.R.S., etc. Edited, with a brief account of the concluding years of his life, by his youngest Daughter, Paulina B. Granville. 2 vols. With a Portrait. Second Edition. Demy 8vo. Cloth, price 32s.

GRAY (Mrs. Russell).
> LISETTE'S VENTURE. A Novel. 2 vols. Crown 8vo. Cloth.

GREEN (T. Bowden).
> FRAGMENTS OF THOUGHT. Dedicated by permission to the Poet Laureate. Crown 8vo. Cloth, price 7s. 6d.

GREENWOOD (James), "The Amateur Casual."
> IN STRANGE COMPANY; or, The Note Book of a Roving Correspondent. Second Edition. Crown 8vo. Cloth, price 6s.

GREY (John), of Dilston.
> JOHN GREY (of Dilston): MEMOIRS. By Josephine E. Butler. New and Cheaper Edition. Crown 8vo. Cloth, price 3s. 6d.

GRIFFITH (Rev. T.), A.M., Prebendary of St. Paul's.
> STUDIES OF THE DIVINE MASTER. Demy 8vo. Cloth, price 12s.

GRIFFITHS (Captain Arthur).
> MEMORIALS OF MILLBANK, AND CHAPTERS IN PRISON HISTORY. With Illustrations. 2 vols. Post 8vo. Cloth, price 21s.
>
> THE QUEEN'S SHILLING. A Novel. 2 vols. Cloth.

GRIMLEY (Rev. H. N.), M.A., Professor of Mathematics in the University of Wales, and Chaplain of Tremadoc Church.
> TREMADOC SERMONS, chiefly on the Spiritual Body, the Unseen World, and the Divine Humanity. Crown 8vo. Cloth, price 7s. 6d.

GRUNER (M. L.)
> STUDIES OF BLAST FURNACE PHENOMENA. Translated by L. D. B. Gordon, F.R.S.E., F.G.S. Demy 8vo. Cloth, price 7s. 6d.

GURNEY (Rev. Archer Thompson).
> WORDS OF FAITH AND CHEER. A Mission of Instruction and Suggestion. Crown 8vo. Cloth, price 6s.
>
> FIRST PRINCIPLES IN CHURCH AND STATE. Demy 8vo. Sewed, price 1s. 6d.

HAECKEL (Professor Ernst), of the University of Jena.
>THE HISTORY OF CREATION. A Popular Account of the Development of the Earth and its Inhabitants, according to the Theories of Kant, Laplace, Lamarck, and Darwin. The Translation revised by Professor E. Ray Lankester, M.A., F.R.S. With Coloured Plates and Genealogical Trees of the various groups of both plants and animals. 2 vols. Post 8vo. Cloth, price 32s.

HARCOURT (Capt. A. F. P.)
>THE SHAKESPEARE ARGOSY: Containing much of the wealth of Shakespeare's Wisdom and Wit, alphabetically arranged and classified. Crown 8vo. Cloth, price 6s.

HAWEIS (Rev. H. R.), M.A.
>SPEECH IN SEASON. Third Edition. Crown 8vo. Cloth, price 9s.
>THOUGHTS FOR THE TIMES. Ninth Edition. Crown 8vo. Cloth, price 7s. 6d.
>UNSECTARIAN FAMILY PRAYERS, for Morning and Evening for a Week, with short selected passages from the Bible. Square crown 8vo. Cloth, price 3s. 6d.

HAWTHORNE (Julian).
>BRESSANT. A Romance. 2 vols. Crown 8vo. Cloth.
>IDOLATRY. A Romance. 2 vols. Crown 8vo. Cloth.

HAWTHORNE (Nathaniel).
>NATHANIEL HAWTHORNE. A Memoir, with Stories now first published in this country. By H. A. Page. Post 8vo. Cloth, price 7s. 6d.
>SEPTIMIUS. A Romance. Second Edition. Crown 8vo. Cloth, price 9s.

HAYMAN (Henry), D.D., late Head Master of Rugby School.
>RUGBY SCHOOL SERMONS. With an Introductory Essay on the Indwelling of the Holy Spirit. Crown 8vo. Cloth, price 7s. 6d.

HEATHERGATE.
>A Story of Scottish Life and Character. By a New Author. 2 vols. Crown 8vo. Cloth.

HELLWALD (Baron F. Von).
>THE RUSSIANS IN CENTRAL ASIA. A Critical Examination, down to the present time, of the Geography and History of Central Asia. Translated by Lieut.-Col. Theodore Wirgman, LL.B. In 1 vol. Large post 8vo. With Map. Cloth, price 12s.

HELVIG (Captain Hugo).
>THE OPERATIONS OF THE BAVARIAN ARMY CORPS. Translated by Captain G. S. Schwabe. With Five large Maps. In 2 vols. Demy 8vo. Cloth, price 24s.

HINTON (James), late Aural Surgeon to Guy's Hospital.
> THE PLACE OF THE PHYSICIAN. Being the Introductory Lecture at Guy's Hospital, 1873-74; to which is added ESSAYS ON THE LAW OF HUMAN LIFE, AND ON THE RELATION BETWEEN ORGANIC AND INORGANIC WORLDS. Second Edition. Crown 8vo. Cloth, price 3s. 6d.
> PHYSIOLOGY FOR PRACTICAL USE. By various Writers. With 50 Illustrations. 2 vols. Second Edition. Crown 8vo. Price 12s. 6d.
> AN ATLAS OF DISEASES OF THE MEMBRANA TYMPANI. With Descriptive Text. Post 8vo. Price £6 6s.
> THE QUESTIONS OF AURAL SURGERY. With Illustrations. 2 vols. Post 8vo. Cloth, price 12s. 6d.

H. J. C.
> THE ART OF FURNISHING. A Popular Treatise on the Principles of Furnishing, based on the Laws of Common Sense, Requirement, and Picturesque Effect. Small crown 8vo. Cloth, price 3s. 6d.

HOCKLEY (W. B.)
> TALES OF THE ZENANA; or, A Nuwab's Leisure Hours. By the Author of "Pandurang Hari." With a Preface by Lord Stanley of Alderley. 2 vols. Crown 8vo. Cloth, price 21s.
> PANDURANG HARI; or, Memoirs of a Hindoo. A Tale of Mahratta Life sixty years ago. With a Preface by Sir H. Bartle E. Frere, G.C.S.I., etc. 2 vols. Crown 8vo. Cloth, price 21s.

HOFFBAUER (Captain).
> THE GERMAN ARTILLERY IN THE BATTLES NEAR METZ. Based on the official reports of the German Artillery. Translated by Capt. E. O. Hollist. With Map and Plans. Demy 8vo. Cloth, price 21s.

HOGAN, M.P.
> A Novel. 3 vols. Crown 8vo. Cloth.

HOLMES (Edmond G. A.)
> POEMS. Fcap. 8vo. Price 5s.

HOLROYD (Major W. R. M.), Bengal Staff Corps, Director of Public Instruction, Punjab.
> TAS-HIL UL KALĀM; or, Hindustani made Easy. Crown 8vo. Cloth, price 5s.

HOPE (Lieut. James).
> IN QUEST OF COOLIES. With Illustrations. Crown 8vo. Cloth, price 6s.

HOOPER (Mrs. G.)
> THE HOUSE OF RABY. With a Frontispiece. Crown 8vo. Cloth, price 3s. 6d.

INTERNATIONAL SCIENTIFIC SERIES (The).

I. **THE FORMS OF WATER IN CLOUDS AND RIVERS, ICE AND GLACIERS.** By J. Tyndall, LL.D., F.R.S. With 14 Illustrations. Sixth Edition. 5s.

II. **PHYSICS AND POLITICS**; or, Thoughts on the Application of the Principles of "Natural Selection" and "Inheritance" to Political Society. By Walter Bagehot. Third Edition. 4s.

III. **FOODS.** By Edward Smith, M.D., LL.B., F.R.S. Profusely Illustrated. Fourth Edition. 5s.

IV. **MIND AND BODY:** The Theories of their Relation. By Alexander Bain, LL.D. With Four Illustrations. Fifth Edition. 4s.

V. **THE STUDY OF SOCIOLOGY.** By Herbert Spencer. Fifth Edition. 5s.

VI. **ON THE CONSERVATION OF ENERGY.** By Balfour Stewart, M.D., LL.D., F.R.S. With 14 Engravings. Third Edition. 5s.

VII. **ANIMAL LOCOMOTION**; or, Walking, Swimming, and Flying By J. B. Pettigrew, M.D., F.R.S. With 130 Illustrations. Second Edition. 5s.

VIII. **RESPONSIBILITY IN MENTAL DISEASE.** By Henry Maudsley, M.D. Second Edition. 5s.

IX. **THE NEW CHEMISTRY.** By Professor J. P. Cooke, of the Harvard University. With 31 Illustrations. Third Edition. 5s.

X. **THE SCIENCE OF LAW.** By Professor Sheldon Amos. Second Edition. 5s.

XI. **ANIMAL MECHANISM.** A Treatise on Terrestrial and Aerial Locomotion. By Professor E. J. Marey. With 117 Illustrations. Second Edition. 5s.

INTERNATIONAL SCIENTIFIC SERIES (The).—*Continued.*

XII. **THE DOCTRINE OF DESCENT AND DARWINISM.** By Professor Oscar Schmidt (Strasburg University). With 26 Illustrations. Third Edition. 5s.

XIII. **THE HISTORY OF THE CONFLICT BETWEEN RELIGION AND SCIENCE.** By Professor J. W. Draper, LL.D. Seventh Edition. 5s.

XIV. **FUNGI;** their Nature, Influences, Uses, etc. By M. C. Cooke, M.A., LL.D. Edited by the Rev. M. J. Berkeley, M.A., F.L.S. With numerous Illustrations. Second Edition. 5s.

XV. **THE CHEMICAL EFFECTS OF LIGHT AND PHOTOGRAPHY.** By Dr. Hermann Vogel (Polytechnic Academy of Berlin). Translation thoroughly revised. With 100 Illustrations. Third Edition. 5s.

XVI. **THE LIFE AND GROWTH OF LANGUAGE.** By William Dwight Whitney, Professor of Sanskrit and Comparative Philology in Yale College, New Haven. Second Edition. 5s.

XVII. **MONEY AND THE MECHANISM OF EXCHANGE.** By W. Stanley Jevons, M.A., F.R.S. Second Edition. 5s.

XVIII. **THE NATURE OF LIGHT:** With a General Account of Physical Optics. By Dr. Eugene Lommel, Professor of Physics in the University of Erlangen. With 188 Illustrations and a table of Spectra in Chromolithography. Second Edition. 5s.

XIX. **ANIMAL PARASITES AND MESSMATES.** By Monsieur Van Beneden, Professor of the University of Louvain, Correspondent of the Institute of France. With 83 Illustrations. Second Edition. 5s.

XX. **FERMENTATION.** By Professor Schützenberger, Director of the Chemical Laboratory at the Sorbonne. Second Edition. 5s.

XXI. **THE FIVE SENSES OF MAN.** By Professor Bernstein, of the University of Halle. With 91 Illustrations. Second Edition. 5s.

INTERNATIONAL SCIENTIFIC SERIES (The).

Forthcoming Volumes.

Prof. W. KINGDON CLIFFORD, M.A. The First Principles of the Exact Sciences explained to the Non-mathematical.

Prof. T. H. HUXLEY, LL.D., F.R.S. Bodily Motion and Consciousness.

Dr. W. B. CARPENTER, LL.D., F.R.S. The Physical Geography of the Sea.

Prof. WILLIAM ODLING, F.R.S. The Old Chemistry viewed from the New Standpoint.

W. LAUDER LINDSAY, M.D., F.R.S.E. Mind in the Lower Animals.

Sir JOHN LUBBOCK, Bart., F.R.S. On Ants and Bees.

Prof. W. T. THISELTON DYER, B.A., B.Sc. Form and Habit in Flowering Plants.

Mr. J. N. LOCKYER, F.R.S. Spectrum Analysis.

Prof. MICHAEL FOSTER, M.D. Protoplasm and the Cell Theory.

H. CHARLTON BASTIAN, M.D., F.R.S. The Brain as an Organ of Mind.

Prof. A. C. RAMSAY, LL.D., F.R.S. Earth Sculpture: Hills, Valleys, Mountains, Plains, Rivers, Lakes; how they were Produced, and how they have been Destroyed.

Prof. RUDOLPH VIRCHOW (Berlin Univ.) Morbid Physiological Action.

Prof. CLAUDE BERNARD. History of the Theories of Life.

Prof. H. SAINTE-CLAIRE DEVILLE. An Introduction to General Chemistry.

Prof. WURTZ. Atoms and the Atomic Theory.

Prof. LACAZE-DUTHIERS. Zoology since Cuvier.

Prof. BERTHELOT. Chemical Synthesis.

INTERNATIONAL SCIENTIFIC SERIES (The).—*Continued.*
(*Forthcoming Volumes.*)

Prof. J. ROSENTHAL. General Physiology of Muscles and Nerves.

Prof. JAMES D. DANA, M.A., LL.D. On Cephalization; or, Head-Characters in the Gradation and Progress of Life.

Prof. S. W. JOHNSON, M.A. On the Nutrition of Plants.

Prof. AUSTIN FLINT, Jr. M.D. The Nervous System, and its Relation to the Bodily Functions.

Prof. FERDINAND COHN (Breslau Univ.) Thallophytes (Algæ, Lichens, Fungi).

Prof. HERMANN (University of Zurich). Respiration.

Prof. LEUCKART (University of Leipsic). Outlines of Animal Organization.

Prof. LIEBREICH (University of Berlin). Outlines of Toxicology.

Prof. KUNDT (University of Strasburg). On Sound.

Prof. REES (University of Erlangen). On Parasitic Plants.

Prof. STEINTHAL (University of Berlin). Outlines of the Science of Language.

P. BERT (Professor of Physiology, Paris). Forms of Life and other Cosmical Conditions.

E. ALGLAVE (Professor of Constitutional and Administrative Law at Douai, and of Political Economy at Lille). The Primitive Elements of Political Constitutions.

P. LORAIN (Professor of Medicine, Paris). Modern Epidemics.

Mons. FREIDEL. The Functions of Organic Chemistry.

Mons. DEBRAY. Precious Metals.

Prof. CORFIELD, M.A., M.D. (Oxon.) Air in its relation to Health.

Prof. A. GIARD. General Embryology.

KING (Alice).
>A CLUSTER OF LIVES. Crown 8vo. Cloth, price 7s. 6d.

KING (Mrs. Hamilton).
>THE DISCIPLES. A New Poem. Second Edition, with some Notes. Crown 8vo. Cloth, price 7s. 6d.
>
>ASPROMONTE, AND OTHER POEMS. Second Edition. Fcap. 8vo. Cloth, price 4s. 6d.

KINGSFORD (Rev. F. W.), M.A., Vicar of St. Thomas's, Stamford Hill; late Chaplain H. E. I. C. (Bengal Presidency).
>HARTHAM CONFERENCES; or, Discussions upon some of the Religious Topics of the Day. "Audi alteram partem." Crown 8vo. Cloth, price 3s. 6d.

KNIGHT (Annette F. C.)
>POEMS. Fcap. 8vo. Cloth, price 5s.

LACORDAIRE (Rev. Père).
>LIFE: Conferences delivered at Toulouse. A New and Cheaper Edition. Crown 8vo. Cloth, price 3s. 6d.

LADY OF LIPARI (The).
>A Poem in Three Cantos. Fcap. 8vo. Cloth, price 5s.

LAURIE (J. S.), of the Inner Temple, Barrister-at-Law; formerly H.M. Inspector of Schools, England; Assistant Royal Commissioner, Ireland; Special Commissioner, African Settlement; Director of Public Instruction, Ceylon.
>EDUCATIONAL COURSE OF SECULAR SCHOOL BOOKS FOR INDIA.
>
>*The following Works are now ready :—*
>
>THE FIRST HINDUSTANI READER. Stiff linen wrapper, price 6d.
>
>THE SECOND HINDUSTANI READER. Stiff linen wrapper, price 6d.
>
>GEOGRAPHY OF INDIA; with Maps and Historical Appendix, tracing the growth of the British Empire in Hindustan. 128 pp. fcap. 8vo. Cloth, price 1s. 6d.

LAYMANN (Captain), Instructor of Tactics at the Military College, Neisse.
>THE FRONTAL ATTACK OF INFANTRY. Translated by Colonel Edward Newdigate. Crown 8vo. Cloth, price 2s. 6d.

L. D. S.
>LETTERS FROM CHINA AND JAPAN. With Illustrated Title-page. Crown 8vo. Cloth, price 7s. 6d.

LEANDER (Richard).
> FANTASTIC STORIES. Translated from the German by Paulina B. Granville. With Eight full-page Illustrations by M. E. Fraser-Tytler. Crown 8vo. Cloth, price 5s.

LEATHES (Rev. Stanley), M.A.
> THE GOSPEL ITS OWN WITNESS. Crown 8vo. Cloth, price 5s.

LEE (Rev. Frederick George), D.C.L.
> THE OTHER WORLD; or, Glimpses of the Supernatural. Being Facts, Records, and Traditions, relating to Dreams, Omens, Miraculous Occurrences, Apparitions, Wraiths, Warnings, Second-sight, Necromancy, Witchcraft, etc. 2 vols. A New Edition. Crown 8vo. Cloth, price 15s.

LEE (Holme).
> HER TITLE OF HONOUR. A Book for Girls. New Edition. With a Frontispiece. Crown 8vo. Cloth, price 5s.

LENOIR (J).
> FAYOUM; or, Artists in Egypt. A Tour with M. Gérome and others. With 13 Illustrations. A New and Cheaper Edition. Crown 8vo. Cloth, price 3s. 6d.

LEWIS (Mary E).
> A RAT WITH THREE TALES. With Four Illustrations by Catherine E. Frere. Cloth, price 5s.

LISTADO (J. T.)
> CIVIL SERVICE. A Novel. 2 vols. Crown 8vo. Cloth.

LOCKER (Frederick).
> LONDON LYRICS. A New and Revised Edition, with Additions and a Portrait of the Author. Crown 8vo. Cloth, elegant, price 7s.

LOMMEL (Dr. Eugene), Professor of Physics in the University of Erlangen.
> THE NATURE OF LIGHT: With a General Account of Physical Optics. Second Edition. With 188 Illustrations and a table of Spectra in Chromolithography. Crown 8vo. Cloth, price 5s.
> Vol. XVIII. of the International Scientific Series.

LORIMER (Peter), D.D.
> JOHN KNOX AND THE CHURCH OF ENGLAND: His work in her Pulpit and his influence upon her Liturgy, Articles, and Parties. Demy 8vo. Cloth, price 12s.

LOVER (Samuel), R.H.A.
> THE LIFE OF SAMUEL LOVER, R.H.A.; Artistic, Literary, and Musical. With Selections from his Unpublished Papers and Correspondence. By Bayle Bernard. 2 vols. With a Portrait. Post 8vo. Cloth, price 21s.

LOWER (Mark Antony), M.A., F.S.A.
> WAYSIDE NOTES IN SCANDINAVIA. Being Notes of Travel in the North of Europe. Crown 8vo. Cloth, price 9s.

LYONS (R. T.), Surgeon-Major, Bengal Army.
> A TREATISE ON RELAPSING FEVER. Post 8vo. Cloth, price 7s. 6d.

MACAULAY (James), M.A., M.D., Edin.
> THE TRUTH ABOUT IRELAND: Tours of Observation in 1872 and 1875. With Remarks on Irish Public Questions. Being a Second Edition of "Ireland in 1872," with a New and Supplementary Preface. Crown 8vo. Cloth, price 3s. 6d.

MAC CARTHY (Denis Florence).
> CALDERON'S DRAMAS. Translated from the Spanish. Post 8vo. Cloth, gilt edges, price 10s.

MAC DONALD (George).
> MALCOLM. A Novel. 3 vols. Second Edition. Crown 8vo. Cloth.
> ST. GEORGE AND ST. MICHAEL. 3 vols. Crown 8vo. Cloth.

MAC KENNA (Stephen J.)
> PLUCKY FELLOWS. A Book for Boys. With Six Illustrations. Second Edition. Crown 8vo. Cloth, price 3s. 6d.
> AT SCHOOL WITH AN OLD DRAGOON. With Six Illustrations. Crown 8vo. Cloth, price 5s.

MACLACHLAN (Archibald Neil Campbell), M.A.
> WILLIAM AUGUSTUS, DUKE OF CUMBERLAND: being a Sketch of his Military Life and Character, chiefly as exhibited in the General Orders of his Royal Highness, 1745—1747. With Illustrations. Post 8vo. Cloth, price 15s.

MAIR (R. S.), M.D., F.R.C.S.E., late Deputy Coroner of Madras.
> THE MEDICAL GUIDE FOR ANGLO-INDIANS. Being a Compendium of Advice to Europeans in India, relating to the Preservation and Regulation of Health. With a Supplement on the Management of Children in India. Crown 8vo. Limp cloth, price 3s. 6d.

MANNING (His Eminence Cardinal).
> ESSAYS ON RELIGION AND LITERATURE. By various Writers. Demy 8vo. Cloth, price 10s. 6d.

MAREY (E. J.)
    ANIMAL MECHANICS. A Treatise on Terrestrial and Aerial Locomotion. With 117 Illustrations. Second Edition. Crown 8vo. Cloth, price 5s.
    Volume XI. of the International Scientific Series.

MARKEWITCH (B.)
    THE NEGLECTED QUESTION. Translated from the Russian, by the Princess Ourousoff, and dedicated by Express Permission to Her Imperial and Royal Highness Marie Alexandrovna, the Duchess of Edinburgh. 2 vols. Crown 8vo. Cloth, price 14s.

MARRIOTT (Maj.-Gen. W. F.), C.S.I.
    A GRAMMAR OF POLITICAL ECONOMY. Crown 8vo. Cloth, price 6s.

MARSHALL (Hamilton).
    THE STORY OF SIR EDWARD'S WIFE. A Novel. Crown 8vo. Cloth, price 10s. 6d.

MASTERMAN (J.)
    HALF-A-DOZEN DAUGHTERS. With a Frontispiece. Crown 8vo. Cloth, price 3s. 6d.

MAUDSLEY (Dr. Henry).
    RESPONSIBILITY IN MENTAL DISEASE. Second Edition. Crown 8vo. Cloth, price 5s.
    Vol. VIII. of the International Scientific Series.

MAUGHAN (William Charles).
    THE ALPS OF ARABIA; or, Travels through Egypt, Sinai, Arabia, and the Holy Land. With Map. A New and Cheaper Edition. Demy 8vo. Cloth, price 5s.

MAURICE (C. Edmund).
    LIVES OF ENGLISH POPULAR LEADERS. No. 1.—STEPHEN LANGTON. Crown 8vo. Cloth, price 7s. 6d.
    No. 2.—TYLER, BALL, and OLDCASTLE. Crown 8vo. Price 7s. 6d.

MEDLEY (Lieut.-Col. J. G.), Royal Engineers.
    AN AUTUMN TOUR IN THE UNITED STATES AND CANADA. Crown 8vo. Cloth, price 5s.

MENZIES (Sutherland).
    MEMOIRS OF DISTINGUISHED WOMEN. Post 8vo. Cloth.

ANNE DE BOURBON.
THE DUCHESS DE LONGUEVILLE.
THE DUCHESS DE CHEVREUSE.
PRINCESS PALATINE.
MADEMOISELLE DE MONTPENSIER.
MADAME DE MONTBAZON.
THE DUCHESS OF PORTSMOUTH.
SARAH JENNINGS.
SARAH, DUCHESS OF MARLBOROUGH.

MICKLETHWAITE (J. T.), F.S.A.
    MODERN PARISH CHURCHES: Their Plan, Design, and Furniture. Crown 8vo. Cloth, price 7s. 6d.

MILNE (James).
    TABLES OF EXCHANGE FOR INDIAN AND CEYLON CURRENCY. Second Edition. Post 8vo. Cloth, price £2 2s.

MIRUS (Major-General von).
    CAVALRY FIELD DUTY. Translated by Major Frank S. Russell, 14th (King's) Hussars. Crown 8vo. Cloth limp, price 7s. 6d.

MIVART (St. George), F.R.S.
    CONTEMPORARY EVOLUTION: Discussing the Theory of Evolution as applied to Science, Art, Religion, and Politics. Post 8vo. Cloth, price 7s. 6d.

MOORE (Rev. Daniel), M.A.
    CHRIST AND HIS CHURCH. By the author of "The Age and the Gospel: Hulsean Lectures," etc. Crown 8vo. Cloth, price 3s. 6d.

MOORE (Rev. Thomas), Vicar of Christ Church, Chesham.
    SERMONETTES: on Synonymous Texts, taken from the Bible and Book of Common Prayer, for the Study, Family Reading, and Private Devotion. Small Crown 8vo. Cloth, price 4s. 6d.

MORELL (J. R.)
    EUCLID SIMPLIFIED IN METHOD AND LANGUAGE. Being a Manual of Geometry. Compiled from the most important French Works, approved by the University of Paris and the Minister of Public Instruction. Fcap. 8vo. Cloth, price 2s. 6d.

MORICE (Rev. F. D.), M.A., Fellow of Queen's College, Oxford.
    THE OLYMPIAN AND PYTHIAN ODES OF PINDAR. A New Translation in English Verse. Crown 8vo. Cloth, price 7s. 6d.

MORLEY (Susan).
    AILEEN FERRERS. A Novel. 2 vols. Crown 8vo. Cloth.
    THROSTLETHWAITE. A Novel. 3 vols. Crown 8vo. Cloth.

MORSE (Edward S.), Ph. D., late Professor of Comparative Anatomy and Zoology in Bowdoin College.
> **FIRST BOOK OF ZOOLOGY.** With numerous Illustrations. Crown 8vo. Cloth, price 5s.

MOSTYN (Sydney).
> **PERPLEXITY.** A Novel. 3 vols. Crown 8vo. Cloth.

MUSGRAVE (Anthony).
> **STUDIES IN POLITICAL ECONOMY.** Crown 8vo. Cloth, price 6s.

MY SISTER ROSALIND.
> A Novel. By the Author of "Christiana North," and "Under the Limes." 2 vols. Cloth.

NAAKÈ (John T.), of the British Museum.
> **SLAVONIC FAIRY TALES.** From Russian, Servian, Polish, and Bohemian Sources. With Four Illustrations. Crown 8vo. Cloth, price 5s.

NEWMAN (John Henry), D.D.
> **CHARACTERISTICS FROM THE WRITINGS OF DR. J. H. NEWMAN.** Being Selections, Personal, Historical, Philosophical, and Religious, from his various Works. Arranged with the Author's personal approval. Second Edition. With Portrait. Crown 8vo. Cloth, price 6s.
> *** A Portrait of the Rev. Dr. J. H. Newman, mounted for framing, can be had, price 2s. 6d.

NEWMAN (Mrs.)
> **TOO LATE.** A Novel. 2 vols. Crown 8vo. Cloth.

NOBLE (James Ashcroft).
> **THE PELICAN PAPERS.** Reminiscences and Remains of a Dweller in the Wilderness. Crown 8vo. Cloth, price 6s.

NORMAN PEOPLE (The).
> **THE NORMAN PEOPLE,** and their Existing Descendants in the British Dominions and the United States of America. Demy 8vo. Cloth, price 21s.

NORRIS (Rev. A.)
> **THE INNER AND OUTER LIFE POEMS.** Fcap. 8vo. Cloth, price 6s.

NOTREGE (John), A.M.
> **THE SPIRITUAL FUNCTION OF A PRESBYTER IN THE CHURCH OF ENGLAND.** Crown 8vo. Cloth, red edges, price 3s. 6d.

ORIENTAL SPORTING MAGAZINE (The).
> A Reprint of the first 5 Volumes, in 2 Volumes. Demy 8vo. Cloth, price 28s.

OUR INCREASING MILITARY DIFFICULTY, and one Way of Meeting it.
    Demy 8vo. Stitched, price 1s.

PAGE (H. A.)
    NATHANIEL HAWTHORNE, A MEMOIR OF, with Stories now first published in this country. Large post 8vo. Cloth, price 7s. 6d.

PAGE (Capt. S. Flood).
    DISCIPLINE AND DRILL. Four Lectures delivered to the London Scottish Rifle Volunteers. Cheaper Edition. Crown 8vo. Price 1s.

PALGRAVE (W. Gifford).
    HERMANN AGHA. An Eastern Narrative. 2 vols. Crown 8vo. Cloth, extra gilt, price 18s.

PANDURANG HARI.
    A Tale of Mahratta Life sixty years ago. With a Preface by Sir H. Bartle E. Frere. 2 vols. Crown 8vo. Cloth, price 21s.

PARKER (Joseph), D.D.
    THE PARACLETE: An Essay on the Personality and Ministry of the Holy Ghost, with some reference to current discussions. Second Edition. Demy 8vo. Cloth, price 12s.

PARR (Harriett).
    ECHOES OF A FAMOUS YEAR. Crown 8vo. Cloth, price 8s. 6d.

PAUL (C. Kegan).
    GOETHE'S FAUST. A New Translation in Rime. Crown 8vo. Cloth, price 6s.
    WILLIAM GODWIN: HIS FRIENDS AND CONTEMPORARIES. With Portraits and Facsimiles of the Handwriting of Godwin and his Wife. 2 vols. Demy 8vo. Cloth, price 28s.
    THE GENIUS OF CHRISTIANITY UNVEILED. Being Essays never before published. Edited, with a Preface, by C. Kegan Paul. Crown 8vo. Price 7s. 6d.

PAYNE (John).
    SONGS OF LIFE AND DEATH. Crown 8vo. Cloth, price 5s.

PAYNE (Professor).
    LECTURES ON EDUCATION. Price 6d. each.
        I. Pestalozzi: the Influence of His Principles and Practice.
        II. Fröbel and the Kindergarten System. Second Edition.
        III. The Science and Art of Education.
        IV. The True Foundation of Science Teaching.

PELLETAN (Eugène).
: THE DESERT PASTOR, JEAN JAROUSSEAU. Translated from the French. By Colonel E. P. De L'Hoste. With a Frontispiece. New Edition. Fcap. 8vo. Cloth, price 3s. 6d.

PENRICE (Major J.), B.A.
: A DICTIONARY AND GLOSSARY OF THE KO-RAN. With copious Grammatical References and Explanations of the Text. 4to. Cloth, price 21s.

PERCEVAL (Rev. P.)
: TAMIL PROVERBS, WITH THEIR ENGLISH TRANSLATION. Containing upwards of Six Thousand Proverbs. Third Edition. Demy 8vo. Sewed, price 9s.

PERRIER (Amelia).
: A WINTER IN MOROCCO. With Four Illustrations. A New and Cheaper Edition. Crown 8vo. Cloth, price 3s. 6d.
: A GOOD MATCH. A Novel. 2 vols. Crown 8vo. Cloth.

PERRY (Rev. S. J.)
: NOTES OF A VOYAGE TO KERGUELEN ISLAND. Royal 8vo. Sewed, price 2s.

PETTIGREW (J. B.), M.D., F.R.S.
: ANIMAL LOCOMOTION; or, Walking, Swimming, and Flying With 130 Illustrations. Second Edition. Crown 8vo. Cloth, price 5s.
: Vol. VII. of the International Scientific Series.

PIGGOT (John), F.S.A, F.R.G.S.
: PERSIA—ANCIENT AND MODERN. Post 8vo. Cloth, price 10s. 6d.

POUSHKIN (Alexander Serguevitch).
: RUSSIAN ROMANCE. Translated from the Tales of Belkin, etc. By Mrs. J. Buchan Telfer (née Mouravieff). Crown 8vo. Cloth, price 7s. 6d.

POWER (Harriet).
: OUR INVALIDS: HOW SHALL WE EMPLOY AND AMUSE THEM? Fcap 8vo. Cloth, price 2s. 6d.

POWLETT (Lieut. Norton), Royal Artillery.
: EASTERN LEGENDS AND STORIES IN ENGLISH VERSE. Crown 8vo. Cloth, price 5s.

PRESBYTER.
: UNFOLDINGS OF CHRISTIAN HOPE. An Essay showing that the Doctrine contained in the Damnatory Clauses of the Creed commonly called Athanasian is unscriptural. Small crown 8vo. Cloth, price 4s. 6d.

PRICE (Prof. Bonamy).
> CURRENCY AND BANKING. Crown 8vo. Cloth, price 6s.

PROCTOR (Richard A.)
> OUR PLACE AMONG INFINITIES. A Series of Essays contrasting our little abode in space and time with the Infinities around us. To which are added Essays on "Astrology," and "The Jewish Sabbath." Second Edition. Crown 8vo. Cloth, price 6s.
>
> THE EXPANSE OF HEAVEN. A Series of Essays on the Wonders of the Firmament. With a Frontispiece. Second Edition. Crown 8vo. Cloth, price 6s.

RANKING (B. Montgomerie).
> STREAMS FROM HIDDEN SOURCES. Crown 8vo. Cloth, price 6s.

READY-MONEY MORTIBOY.
> A Matter-of-Fact Story. With Frontispiece. Crown 8vo. Cloth, price 3s. 6d.

REANEY (Mrs. G. S.)
> WAKING AND WORKING; OR, FROM GIRLHOOD TO WOMANHOOD. With a Frontispiece. Crown 8vo. Cloth, price 5s.
>
> SUNBEAM WILLIE, AND OTHER STORIES, for Home Reading and Cottage Meetings. 3 Illustrations. Small square, uniform with "Lost Gip," etc. Price 1s. 6d.

REGINALD BRAMBLE.
> A Cynic of the Nineteenth Century. An Autobiography. Crown 8vo. Cloth, price 10s. 6d.

REID (T. Wemyss).
> CABINET PORTRAITS. Biographical Sketches of Statesmen of the Day. Crown 8vo. Cloth, price 7s. 6d.

RHOADES (James).
> TIMOLEON. A Dramatic Poem. Fcap. 8vo. Cloth, price 5s.

RIBOT (Professor Th.)
> CONTEMPORARY ENGLISH PSYCHOLOGY. Second Edition. A revised and corrected translation from the latest French Edition. Large post 8vo. Cloth, price 9s.
>
> HEREDITY: A Psychological Study on its Phenomena, its Laws, its Causes, and its Consequences. Large crown 8vo. Cloth, price 9s.

ROBERTSON (The Late Rev. F. W.), M.A.

THE LATE REV. F. W. ROBERTSON, M.A., LIFE AND LETTERS OF. Edited by the Rev. Stopford Brooke, M.A., Chaplain in Ordinary to the Queen.

I. 2 vols., uniform with the Sermons. With Steel Portrait. Crown 8vo. Cloth, price 7s. 6d.
II. Library Edition, in Demy 8vo. with Two Steel Portraits. Cloth, price 12s.
III. A Popular Edition, in 1 vol. Crown 8vo. Cloth, price 6s.

New and Cheaper Editions :—

SERMONS.
    First Series. Small crown 8vo. Cloth, price 3s. 6d.
    Second Series. Small crown 8vo. Cloth, price 3s. 6d.
    Third Series. Small crown 8vo. Cloth, price 3s. 6d.
    Fourth Series. Small crown 8vo. Cloth, price 3s. 6d.

EXPOSITORY LECTURES ON ST. PAUL'S EPISTLE TO THE CORINTHIANS. Small crown 8vo. Cloth, price 5s.

LECTURES AND ADDRESSES, with other literary remains. A New Edition. Crown 8vo. Cloth, price 5s.

AN ANALYSIS OF MR. TENNYSON'S "IN MEMORIAM." (Dedicated by Permission to the Poet-Laureate.) Fcap. 8vo. Cloth, price 2s.

THE EDUCATION OF THE HUMAN RACE. Translated from the German of Gotthold Ephraim Lessing. Fcap. 8vo. Cloth, price 2s. 6d.

The above Works can also be had bound in half-morocco.

*** A Portrait of the late Rev. F. W. Robertson, mounted for framing, can be had, price 2s. 6d.

ROSS (Mrs. Ellen), ("Nelsie Brook.")

DADDY'S PET. A Sketch from Humble Life. Uniform with "Lost Gip." With Six Illustrations. Square crown 8vo. Cloth, price 1s.

ROXBURGHE LOTHIAN.

DANTE AND BEATRICE FROM 1282 TO 1290. A Romance. 2 vols. Post 8vo. Cloth, price 24s.

RUSSELL (William Clark).

MEMOIRS OF MRS. LETITIA BOOTHBY. Crown 8vo. Cloth, price 7s. 6d.

RUSSELL (E. R.)

IRVING AS HAMLET. Second Edition. Demy 8vo. Sewed, price 1s.

SADLER (S. W.), R.N., Author of "Marshall Vavasour."
> THE AFRICAN CRUISER. A Midshipman's Adventures on the West Coast. A Book for Boys. With Three Illustrations. Second Edition. Crown 8vo. Cloth, price 3s. 6d.

SAMAROW (Gregor).
> FOR SCEPTRE AND CROWN. A Romance of the Present Time. Translated by Fanny Wormald. 2 vols. Crown 8vo. Cloth, price 15s.

SAUNDERS (Katherine).
> THE HIGH MILLS. A Novel. 3 vols. Crown 8vo. Cloth.
>
> GIDEON'S ROCK, and other Stories. Crown 8vo. Cloth, price 6s.
>
> JOAN MERRYWEATHER, and other Stories. Crown 8vo. Cloth, price 6s.
>
> MARGARET AND ELIZABETH. A Story of the Sea. Crown 8vo. Cloth, price 6s.

SAUNDERS (John).
> ISRAEL MORT, OVERMAN. A Story of the Mine. 3 vols. Crown 8vo.
>
> HIRELL. With Frontispiece. Crown 8vo. Cloth, price 3s. 6d.
>
> ABEL DRAKE'S WIFE. With Frontispiece. Crown 8vo. Cloth, price 3s. 6d.

SCHELL (Major von).
> THE OPERATIONS OF THE FIRST ARMY UNDER GEN. VON GOEBEN. Translated by Col. C. H. von Wright. Four Maps. Demy 8vo. Cloth, price 9s.
>
> THE OPERATIONS OF THE FIRST ARMY UNDER GEN. VON STEINMETZ. Translated by Captain E. O. Hollist. Demy 8vo. Cloth, price 10s. 6d.

SCHERFF (Major W. von).
> STUDIES IN THE NEW INFANTRY TACTICS. Parts I. and II. Translated from the German by Colonel Lumley Graham. Demy 8vo. Cloth, price 7s. 6d.

SCHMIDT (Prof. Oscar), Strasburg University.
> THE DOCTRINE OF DESCENT AND DARWINISM. With 26 Illustrations. Third Edition. Crown 8vo. Cloth, price 5s. Vol. XII. of the International Scientific Series.

SCHÜTZENBERGER (Prof. F.), Director of the Chemical Laboratory at the Sorbonne.
> FERMENTATION. With numerous Illustrations. Crown 8vo. Cloth, price 5s.
> Vol. XX. of the International Scientific Series.

SCOTT (Patrick).
THE DREAM AND THE DEED, and other Poems. Fcap. 8vo. Cloth, price 5s.

SCOTT (W. T.)
ANTIQUITIES OF GREAT DUNMOW. Crown 8vo. Cloth, price 5s.

SEEKING HIS FORTUNE, and other Stories.
With Four Illustrations. Crown 8vo. Cloth, price 3s. 6d.

SENIOR (Nassau William).
ALEXIS DE TOCQUEVILLE. Correspondence and Conversations with Nassau W. Senior, from 1833 to 1859. Edited by M. C. M. Simpson. 2 vols. Large post 8vo. Cloth, price 21s.
JOURNALS KEPT IN FRANCE AND ITALY. From 1848 to 1852. With a Sketch of the Revolution of 1848. Edited by his Daughter, M. C. M. Simpson. 2 vols. Post 8vo. Cloth, price 24s.

SEVEN AUTUMN LEAVES FROM FAIRYLAND.
Illustrated with Nine Etchings. Square crown 8vo. Cloth, price 3s. 6d.

SEYD (Ernest), F.S.S.
THE FALL IN THE PRICE OF SILVER. Its Causes, its Consequences, and their Possible Avoidance, with Special Reference to India. Demy 8vo. Sewed, price 2s. 6d.

SHADWELL (Major-General), C.B.
MOUNTAIN WARFARE. Illustrated by the Campaign of 1799 in Switzerland. Being a Translation of the Swiss Narrative compiled from the Works of the Archduke Charles, Jomini, and others. Also of Notes by General H. Dufour on the Campaign of the Valtelline in 1635. With Appendix, Maps, and Introductory Remarks. Demy 8vo. Cloth, price 16s.

SHELDON (Philip).
WOMAN'S A RIDDLE; or, Baby Warmstrey. A Novel. 3 vols. Crown 8vo. Cloth.

SHERMAN (Gen. W. T.)
MEMOIRS OF GEN. W. T. SHERMAN, Commander of the Federal Forces in the American Civil War. By Himself. 2 vols. With Map. Demy 8vo. Cloth, price 24s. *Copyright English Edition.*

SHELLEY (Lady).
SHELLEY MEMORIALS FROM AUTHENTIC SOURCES. With (now first printed) an Essay on Christianity by Percy Bysshe Shelley. With Portrait. Third Edition. Crown 8vo. Cloth, price 5s.

SHIPLEY (Rev. Orby), M.A.
   STUDIES IN MODERN PROBLEMS. By various Writers. 2 vols. Crown 8vo. Cloth, price 5s. each.

   CONTENTS.—VOL. I.

   Sacramental Confession.
   Abolition of the Thirty-nine Articles. Part I.
   The Sanctity of Marriage.
   Creation and Modern Science.
   Retreats for Persons Living in the World.
   Catholic and Protestant.
   The Bishops on Confession in the Church of England.

   CONTENTS.—VOL. II.

   Some Principles of Christian Ceremonial.
   A Layman's View of Confession of Sin to a Priest. Parts I. and II.
   Reservation of the Blessed Sacrament.
   Missions and Preaching Orders.
   Abolition of the Thirty-nine Articles. Part II.
   The First Liturgy of Edward VI. and our own office contrasted and compared.

SMEDLEY (M. B.)
   BOARDING-OUT AND PAUPER SCHOOLS FOR GIRLS. Crown 8vo. Cloth, price 3s. 6d.

SMITH (Edward), M.D., LL.B., F.R.S.
   HEALTH AND DISEASE, as influenced by the Daily, Seasonal, and other Cyclical Changes in the Human System. A New Edition. Post 8vo. Cloth, price 7s. 6d.
   FOODS. Profusely Illustrated. Fourth Edition. Crown 8vo. Cloth, price 5s.
   Vol. III. of the International Scientific Series.
   PRACTICAL DIETARY FOR FAMILIES, SCHOOLS, AND THE LABOURING CLASSES. A New Edition. Post 8vo. Cloth, price 3s. 6d.
   CONSUMPTION IN ITS EARLY AND REMEDIABLE STAGES. A New Edition. Crown 8vo. Cloth, price 6s.

SMITH (Hubert).
   TENT LIFE WITH ENGLISH GIPSIES IN NORWAY. With Five full-page Engravings and Thirty-one smaller Illustrations by Whymper and others, and Map of the Country showing Routes. Second Edition. Revised and Corrected. Post 8vo. Cloth, price 21s.

SONGS FOR MUSIC.
   By Four Friends. Square crown 8vo. Cloth, price 5s.
   Containing Songs by Reginald A. Gatty, Stephen H. Gatty, Greville J. Chester, and Juliana H. Ewing.

SOME TIME IN IRELAND.
  A Recollection. Crown 8vo. Cloth, price 7s. 6d.

SONGS OF TWO WORLDS.
  SONGS OF TWO WORLDS. By a New Writer. First Series. Second Edition. Fcap. 8vo. Cloth, price 5s.
  SONGS OF TWO WORLDS. By a New Writer. Second Series. Second Edition. Fcap. 8vo. Cloth, price 5s.
  SONGS OF TWO WORLDS. By a New Writer. Third Series. Second Edition. Fcap. 8vo. Cloth, price 5s.
  THE EPIC OF HADES. By the Author of "Songs of Two Worlds." Fcap. 8vo. Cloth, price 5s.

SPENCER (HERBERT).
  THE STUDY OF SOCIOLOGY. Fifth Edition. Crown 8vo. Cloth, price 5s.
  Vol. V. of the International Scientific Series.

SPICER (Henry),
  OTHO'S DEATH WAGER. A Dark Page of History Illustrated. In Five Acts. Fcap. 8vo. Cloth, price 5s.

STEVENSON (Rev. W. Fleming),
  HYMNS FOR THE CHURCH AND HOME. Selected and Edited by the Rev. W. Fleming Stevenson.
  The most complete Hymn Book published.
  The Hymn Book consists of Three Parts:—I. For Public Worship.—II. For Family and Private Worship.—III. For Children.
  ₂ Published in various forms and prices, the latter ranging from 8d. to 6s. Lists and full particulars will be furnished on application to the Publishers.

STEWART (Professor Balfour).
  ON THE CONSERVATION OF ENERGY. Third Edition. With Fourteen Engravings. Crown 8vo. Cloth, price 5s.
  Vol. VI. of the International Scientific Series.

STONEHEWER (Agnes).
  MONACELLA: A Legend of North Wales. A Poem. Fcap. 8vo. Cloth, price 3s. 6d.

STRETTON (Hesba). Author of "Jessica's First Prayer."
  THE CREW OF THE DOLPHIN. Illustrated. Square crown 8vo. Cloth, price 1s. 6d.
  CASSY. Twenty-seventh Thousand. With Six Illustrations. Square crown 8vo. Cloth, price 1s. 6d.
  THE KING'S SERVANTS. Thirty-third Thousand. With Eight Illustrations. Square crown 8vo. Cloth, price 1s. 6d.

STRETTON (Hesba). Author of " Jessica's First Prayer."
    LOST GIP. Forty-seventh Thousand. With Six Illustrations. Square crown 8vo. Cloth, price 1s. 6d.
    *** *Also a handsomely-bound Edition, with Twelve Illustrations, price 2s. 6d.*
    THE WONDERFUL LIFE. Ninth Thousand. Fcap. 8vo. Cloth, price 2s. 6d.
    FRIENDS TILL DEATH. With Frontispiece. Limp cloth, price 6d.
    TWO CHRISTMAS STORIES. With Frontispiece. Limp cloth, price 6d.
    MICHEL LORIO'S CROSS, AND LEFT ALONE. With Frontispiece. Limp cloth, price 6d.
    OLD TRANSOME. With Frontispiece. Limp cloth, price 6d.
    THE WORTH OF A BABY, AND HOW APPLE-TREE COURT WAS WON. With Frontispiece. Limp cloth, price 6d.
    HESTER MORLEY'S PROMISE. 3 vols. Crown 8vo. Cloth.
    THE DOCTOR'S DILEMMA. 3 vols. Crown 8vo. Cloth.

SULLY (James).
    SENSATION AND INTUITION. Demy 8vo. Cloth, price 10s. 6d.

TALES OF THE ZENANA.
    By the Author of "Pandurang Hari." 2 vols. Crown 8vo. Cloth, price 21s.

TAYLOR (Rev. J. W. Augustus), M.A.
    POEMS. Fcap. 8vo. Cloth, price 5s.

TAYLOR (Sir Henry).
    EDWIN THE FAIR AND ISAAC COMNENUS. Fcap. 8vo. Cloth, price 3s. 6d.
    A SICILIAN SUMMER AND OTHER POEMS. Fcap. 8vo. Cloth, price 3s. 6d.
    PHILIP VAN ARTEVELDE. A Dramatic Poem. Fcap. 8vo. Cloth, price 5s.

TAYLOR (Colonel Meadows), C.S.I., M.R.I.A.
    SEETA. A Novel. 3 vols. Crown 8vo. Cloth.
    THE CONFESSIONS OF A THUG. Crown 8vo. Cloth, price 6s.
    TARA: a Mahratta Tale. Crown 8vo. Cloth, price 6s.

THOMAS (Moy).
    A FIGHT FOR LIFE. With Frontispiece. Crown 8vo. Cloth, price 3s. 6d.

THOMSON (J. T.), F.R.G.S.
    HAKAYIT ABDULLA. The Autobiography of a Malay Munshi, between the years 1808 and 1843. Demy 8vo. Cloth, price 12s.

TENNYSON (Alfred).
QUEEN MARY. A Drama. New Edition. Crown 8vo. Cloth, price 6s.

TENNYSON'S (Alfred) Works. Cabinet Edition. Ten Volumes. Each with Portrait. Fcap. 8vo. Cloth, price 2s. 6d.
CABINET EDITION. 10 vols. Complete in handsome Ornamental Case. Price 28s.

TENNYSON'S (Alfred) Works. Author's Edition. Complete in Five Volumes. Post 8vo. Cloth gilt; or half-morocco, Roxburgh style.
VOL. I. EARLY POEMS, and ENGLISH IDYLLS. Price 6s.; Roxburgh, 7s. 6d.
VOL. II. LOCKSLEY HALL, LUCRETIUS, and other Poems. Price 6s.; Roxburgh, 7s. 6d.
VOL. III. THE IDYLLS OF THE KING (*Complete*). Price 7s. 6d.; Roxburgh, 9s.
VOL. IV. THE PRINCESS, and MAUD. Price 6s.; Roxburgh, 7s. 6d.
VOL. V. ENOCH ARDEN, and IN MEMORIAM. Price 6s.: Roxburgh, 7s. 6d.

TENNYSON'S IDYLLS OF THE KING, and other Poems. Illustrated by Julia Margaret Cameron. 1 vol. Folio. Half-bound morocco, cloth sides. Six Guineas.

TENNYSON'S (Alfred) Works. Original Editions.
POEMS. Small 8vo. Cloth, price 6s.
MAUD, and other Poems. Small 8vo. Cloth, price 3s. 6d.
THE PRINCESS. Small 8vo. Cloth, price 3s. 6d.
IDYLLS OF THE KING. Small 8vo. Cloth, price 5s.
IDYLLS OF THE KING. Collected. Small 8vo. Cloth, price 6s.
THE HOLY GRAIL, and other Poems. Small 8vo. Cloth, price 4s. 6d.
GARETH AND LYNETTE. Small 8vo. Cloth, price 3s.
ENOCH ARDEN, etc. Small 8vo. Cloth, price 3s. 6d.
SELECTIONS FROM THE ABOVE WORKS. Square 8vo. Cloth, price 3s. 6d. Cloth gilt, extra, price 4s.
SONGS FROM THE ABOVE WORKS. Square 8vo. Cloth extra, price 3s. 6d.
IN MEMORIAM. Small 8vo. Cloth, price 4s.
LIBRARY EDITION. In 6 vols. Demy 8vo. Cloth, price 10s. 6d. each.
POCKET VOLUME EDITION. 11 vols. In neat case, 31s. 6d.
Ditto, ditto. Extra cloth gilt, in case, 35s.
POEMS. Illustrated Edition. 4to. Cloth, price 25s.

THOMASINA.
>   A Novel. 2 vols. Crown 8vo.

THOMPSON (A. C.)
>   PRELUDES. A Volume of Poems. Illustrated by Elizabeth Thompson (Painter of "The Roll Call"). 8vo. Cloth, price 7s. 6d.

THOMPSON (Rev. A. S.), British Chaplain at St. Petersburg.
>   HOME WORDS FOR WANDERERS. A Volume of Sermons. Crown 8vo. Cloth, price 6s.

THOUGHTS IN VERSE.
>   Small crown 8vo. Cloth, price 1s. 6d.

THRING (Rev. Godfrey), B.A.
>   HYMNS AND SACRED LYRICS. Fcap. 8vo. Cloth, price 5s.

TODD (Herbert), M.A.
>   ARVAN; or, The Story of the Sword. A Poem. Crown 8vo. Cloth, price 7s. 6d.

TRAHERNE (Mrs. Arthur).
>   THE ROMANTIC ANNALS OF A NAVAL FAMILY. A New and Cheaper Edition. Crown 8vo. Cloth, price 5s.

TRAVERS (Mar.)
>   THE SPINSTERS OF BLATCHINGTON. A Novel. 2 vols. Crown 8vo. Cloth.

TREVANDRUM OBSERVATIONS.
>   OBSERVATIONS OF MAGNETIC DECLINATION MADE AT TREVANDRUM AND AGUSTIA MALLEY in the Observatories of his Highness the Maharajah of Travancore, G.C.S.I., in the Years 1852 to 1860. Being Trevandrum Magnetical Observations, Volume I. Discussed and Edited by John Allan Brown, F.R.S., late Director of the Observatories. With an Appendix. Imp. 4to. Cloth, price £3 3s.
>   *** *The Appendix, containing Reports on the Observatories and on the Public Museum, Public Park, and Gardens at Trevandrum, pp.* xii.-116, *may be had separately, price* 21s.

TURNER (Rev. Charles).
>   SONNETS, LYRICS, AND TRANSLATIONS. Crown 8vo. Cloth, price 4s. 6d.

TYNDALL (J.), LL.D., F.R.S.
>   THE FORMS OF WATER IN CLOUDS AND RIVERS, ICE AND GLACIERS. With Twenty-six Illustrations. Sixth Edition. Crown 8vo. Cloth, price 5s.
>   Vol. I. of the International Scientific Series.

UMBRA OXONIENSIS.
>   RESULTS OF THE EXPOSTULATION OF THE RIGHT
>   HONOURABLE W. E. GLADSTONE, in their Relation to the
>   Unity of Roman Catholicism. Large fcap. 8vo. Cloth, price 5s.

UPTON (Roger D.), Captain late 9th Royal Lancers.
>   NEWMARKET AND ARABIA. An Examination of the
>   Descent of Racers and Coursers. With Pedigrees and Frontispiece. Post 8vo. Cloth, price 9s.

VAMBERY (Prof. Arminius), of the University of Pesth.
>   BOKHARA: Its History and Conquest. Demy 8vo. Cloth, price 18s.

VAN BENEDEN (Monsieur), Professor of the University of Louvain, Correspondent of the Institute of France.
>   ANIMAL PARASITES AND MESSMATES. With 83 Illustrations. Second Edition. Cloth, price 5s.
>   Vol. XIX. of the International Scientific Series.

VANESSA.
>   By the Author of "Thomasina," etc. A Novel. 2 vols. Second Edition. Crown 8vo. Cloth.

VAUGHAN (Rev. C. J.), D.D.
>   WORDS OF HOPE FROM THE PULPIT OF THE TEMPLE
>   CHURCH. Third Edition. Crown 8vo. Cloth, price 5s.
>   THE SOLIDITY OF TRUE RELIGION, and other Sermons
>   Preached in London during the Election and Mission Week, February, 1874. Crown 8vo. Cloth, price 3s. 6d.
>   FORGET THINE OWN PEOPLE. An Appeal for Missions.
>   Crown 8vo. Cloth, price 3s. 6d.
>   THE YOUNG LIFE EQUIPPING ITSELF FOR GOD'S SERVICE. Being Four Sermons Preached before the University of Cambridge, in November, 1872. Fourth Edition. Crown 8vo. Cloth, price 3s. 6d.

VINCENT (Capt. C. E. H.), late Royal Welsh Fusiliers.
>   ELEMENTARY MILITARY GEOGRAPHY, RECONNOITRING,
>   AND SKETCHING. Compiled for Non-Commissioned Officers
>   and Soldiers of all Arms. Square crown 8vo. Cloth, price 2s. 6d.
>   RUSSIA'S ADVANCE EASTWARD. Based on the Official
>   Reports of Lieutenant Hugo Stumm, German Military Attaché
>   to the Khivan Expedition. With Map. Crown 8vo. Cloth, price 6s.

VIZCAYA; or, Life in the Land of the Carlists at the
>   Outbreak of the Insurrection, with some Account of the Iron Mines
>   and other Characteristics of the Country. With a Map and Eight
>   Illustrations. Crown 8vo. Cloth, price 9s.

VOGEL (Prof.), Polytechnic Academy of Berlin.
>   THE CHEMICAL EFFECTS OF LIGHT AND PHOTOGRAPHY, in their application to Art, Science, and Industry. The translation thoroughly revised. With 100 Illustrations, including some beautiful Specimens of Photography. Third Edition. Crown 8vo. Cloth, price 5s.
>   Vol. XV. of the International Scientific Series.

VYNER (Lady Mary).
>   EVERY DAY A PORTION. Adapted from the Bible and the Prayer Book, for the Private Devotions of those living in Widowhood. Collected and Edited by Lady Mary Vyner. Square crown 8vo. Cloth extra, price 5s.

WAITING FOR TIDINGS.
>   By the Author of "White and Black." 3 vols. Crown 8vo. Cloth.

WARTENSLEBEN (Count Hermann von), Colonel in the Prussian General Staff.
>   THE OPERATIONS OF THE SOUTH ARMY IN JANUARY AND FEBRUARY, 1871. Compiled from the Official War Documents of the Head-quarters of the Southern Army. Translated by Colonel C. H. von Wright. With Maps. Demy 8vo. Cloth, price 6s.
>   THE OPERATIONS OF THE FIRST ARMY UNDER GEN. VON MANTEUFFEL. Translated by Colonel C. H. von Wright. Uniform with the above. Demy 8vo. Cloth, price 9s.

WEDMORE (Frederick).
>   TWO GIRLS. 2 vols. Crown 8vo. Cloth.

WELLS (Captain John C.), R.N.
>   SPITZBERGEN—THE GATEWAY TO THE POLYNIA; or, A Voyage to Spitzbergen. With numerous Illustrations by Whymper and others, and Map. New and Cheaper Edition. Demy 8vo. Cloth, price 6s.

WETMORE (W. S.).
>   COMMERCIAL TELEGRAPHIC CODE. Post 4to. Boards, price 42s.

WHAT 'TIS TO LOVE.
>   By the Author of "Flora Adair," "The Value of Fostertown." 3 vols. Crown 8vo. Cloth.

WHITAKER (Florence).
>   CHRISTY'S INHERITANCE: A London Story. Illustrated. Royal 32mo. Cloth, price 1s. 6d.

WHITE (Captain F. B. P.)
> THE SUBSTANTIVE SENIORITY ARMY LIST—MAJORS AND CAPTAINS. 8vo. Sewed, price 2s. 6d.

WHITNEY (William Dwight). Professor of Sanskrit and Comparative Philology in Yale College, New Haven.
> THE LIFE AND GROWTH OF LANGUAGE. Second Edition. Crown 8vo. Cloth, price 5s. *Copyright Edition.*
> Vol. XVI. of the International Scientific Series.

WHITTLE (J. Lowry), A.M., Trin. Coll., Dublin.
> CATHOLICISM AND THE VATICAN. With a Narrative of the Old Catholic Congress at Munich. Second Edition. Crown 8vo. Cloth, price 4s. 6d.

WILBERFORCE (Henry W.)
> THE CHURCH AND THE EMPIRES. Historical Periods. Preceded by a Memoir of the Author by John Henry Newman, D.D., of the Oratory. With Portrait. Post 8vo. Cloth, price 10s. 6d.

WILKINSON (T. Lean).
> SHORT LECTURES ON THE LAND LAWS. Delivered before the Working Men's College. Crown 8vo. Limp cloth, price 2s.

WILLIAMS (A. Lukyn), Jesus College, Cambridge.
> FAMINES IN INDIA; their Causes and Possible Prevention. The Essay for the Le Bas Prize, 1875. Demy 8vo. Price 5s.

WILLIAMS (Rev. Rowland), D.D.
> LIFE AND LETTERS OF ROWLAND WILLIAMS, D.D., with Selections from his Note-books. Edited by Mrs. Rowland Williams. With a Photographic Portrait. 2 vols. Large post 8vo. Cloth, price 24s.
>
> THE PSALMS, LITANIES, COUNSELS, AND COLLECTS FOR DEVOUT PERSONS. Edited by his Widow. New and Popular Edition. Crown 8vo. Cloth, price 3s. 6d.

WILLOUGHBY (The Hon. Mrs.)
> ON THE NORTH WIND—THISTLEDOWN. A Volume of Poems. Elegantly bound. Small crown 8vo. Cloth, price 7s. 6d.

WILSON (H. Schütz).
> STUDIES AND ROMANCES. Crown 8vo. Cloth, price 7s. 6d.

WILSON (Lieutenant-Colonel C. Townshend).
> JAMES THE SECOND AND THE DUKE OF BERWICK. Demy 8vo. Cloth, price 12s. 6d.

WINTERBOTHAM (Rev. R.), M.A., B.Sc.
> SERMONS AND EXPOSITIONS. Crown 8vo. Cloth, price 7s. 6d.

WORNOVITS (Captain Illia).
: AUSTRIAN CAVALRY EXERCISE. Translated by Captain W. S. Cooke. Crown 8vo. Cloth, price 7s.

WOOD (C. F.)
: A YACHTING CRUISE IN THE SOUTH SEAS. With Six Photographic Illustrations. Demy 8vo. Cloth, price 7s. 6d.

WRIGHT (Rev. W.), of Stoke Bishop, Bristol.
: MAN AND ANIMALS: A Sermon. Crown 8vo. Stitched in wrapper, price 1s.
: WAITING FOR THE LIGHT, AND OTHER SERMONS. Crown 8vo. Cloth, price 6s.

WYLD (R. S.), F.R.S.E.
: THE PHYSICS AND PHILOSOPHY OF THE SENSES; or, The Mental and the Physical in their Mutual Relation. Illustrated by several Plates. Demy 8vo. Cloth, price 16s.

YONGE (C. D.), Regius Professor, Queen's College, Belfast.
: HISTORY OF THE ENGLISH REVOLUTION OF 1688. Crown 8vo. Cloth, price 6s.

YORKE (Stephen), Author of "Tales of the North Riding."
: CLEVEDEN. A Novel. 2 vols. Crown 8vo. Cloth.

YOUMANS (Eliza A.)
: AN ESSAY ON THE CULTURE OF THE OBSERVING POWERS OF CHILDREN, especially in connection with the Study of Botany. Edited, with Notes and a Supplement, by Joseph Payne, F.C.P., Author of "Lectures on the Science and Art of Education," etc. Crown 8vo. Cloth, price 2s. 6d.
: FIRST BOOK OF BOTANY. Designed to cultivate the Observing Powers of Children. With 300 Engravings. New and Enlarged Edition. Crown 8vo. Cloth, price 5s.

YOUMANS (Edward L.), M.D.
: A CLASS BOOK OF CHEMISTRY, on the Basis of the new System. With 200 Illustrations. Crown 8vo. Cloth, price 5s.

ZIMMERN (Helen).
: STORIES IN PRECIOUS STONES. With Six Illustrations. Third Edition. Crown 8vo. Cloth, price 5s.

www.ingramcontent.com/pod-product-compliance
Lightning Source LLC
Chambersburg PA
CBHW020827230426
43666CB00007B/1133